Membrane Proteins
Isolation and Characterization

Edited by

Angelo Azzi, Lanfranco Masotti
Arnaldo Vecli

With Contributions by

A. Azzi, M. Baltscheffsky, M. R. Block, R. Bolli, F. Boulay
G. Brandolin, C. Broger, N. Capitanio, E. Casali, P. Cavatorta
V. Comaschi, G. Farruggia, M. B. Ferrari, L. Franzoni, N. Gesmundo
J. Johns, R. Lüthy, C. W. Mahoney, L. Masotti, C. Montecucco
M. Müller, K. A. Nałęcz, M. J. Nałęcz, P. Nyrén, S. Papa, G. Sartor
A. Spisni, A. G. Szabo, A. Szewczyk, P. V. Vignais, J. Von Berger
L. Wojtczak

With 58 Figures

Springer-Verlag
Berlin Heidelberg New York
London Paris Tokyo

Professor Dr. ANGELO AZZI
Institut für Biochemie und Molekularbiologie
Universität Bern
Bühlstraße 28
CH-3012 Bern

Professor Dr. LANFRANCO MASOTTI
Istituto di Chimica Biologica
Università di Parma
Via Gramsci, 14
I-43100 Parma

Professor Dr. ARNALDO VECLI
Istituto di Fisica
Università di Parma
Via M. D'Azeglio, 85
I-43100 Parma

ISBN 3-540-17014-6 Springer-Verlag Berlin Heidelberg New York
ISBN 0-387-17014-6 Springer-Verlag New York Berlin Heidelberg

Library of Congress Cataloging-in-Publication Data. Membrane proteins. Papers from a course sponsored by the Federation of European Biochemical Societies and the Italian Research Council. Bibliography: p. Includes index. 1. Membrane proteins–Analysis–Congresses. I. Azzi, A. (Angelo). II. Federation of European Biochemical Societies. III. Consigilio nazionale delle ricerche (Italy) [DNLM: 1. Membrane Proteins–analysis–congresses. 2. Membrane Proteins–isolation & purification–congresses. QR 55 A5342] QP552.M44M47 1986 574.87'5 86-25988

Printing and bookbinding: Druckhaus Beltz, Hemsbach/Bergstr.
2131/3130-543210

Preface

This volume is the third of a series on Membrane Proteins and, like the preceding manuals, is the result of an International Advanced Course entitled *Isolation and Characterization of Membrane Proteins: Biochemical and Biophysical Aspects* sponsored by the Federation of European Biochemical Societies (FEBS) and the Italian Research Council (CNR).

The success of the course and the continuous development in the field of membrane biology has prompted me to publish also in this case the protocols of the experiments which were carried out by the students.

The students have been able not only to perform the experiments published in this manual without help from the instructors, but also to suggest improvements, which have been incorporated in the published version.

Care has been taken in making the planning and the execution of the experiments as simple as possible, by listing in detail all the necessary pieces of equipment, test tubes, pipettes, chemicals, etc. At the same time the introduction and the "philosophy" have been limited to the essential, as also the references, only those having been listed which may help in a better understanding of the principles and of the biological background of a given experiment.

Like the previous manuals, this one can be useful in the research laboratory where new experiments can be based on carefully checked routine protocols. It can be useful also in a laboratory of clinical analysis, where scientists may be looking for new, simple methodologies suitable to be applied to the field of human clinical testing. Finally, it can be used as a textbook for a practical course at graduate level of biochemistry and of biophysics.

October 1986

ANGELO AZZI
LANFRANCO MASOTTI
ARNALDO VECLI

Contents

II. Lipid-Protein Interaction

III. Protein Modification

IV. Protein Reconstitution

Contributors

Azzi, A., Institut für Biochemie und Molekularbiologie der Universität Bern, Bühlstraße 28, CH-3012 Bern

Baltscheffsky, M., Department of Biochemistry, Arrhenius Laboratory, University of Stockholm, S-106 91 Stockholm

Block, M.R., Laboratoire de Biochimie, Centre d'Etudes Nucléaires 85 X, F-38041 Grenoble Cedex

Bolli, R., Institut für Biochemie und Molekularbiologie der Universität Bern, Bühlstraße 28, CH-3012 Bern

Boulay, F., Laboratoire de Biochimie, Centre d'Etudes Nucléaires 85 X, F-38041 Grenoble Cedex

Brandolin, G., Laboratoire de Biochimie, Centre d'Etudes Nucléaires 85 X, F-38041 Grenoble Cedex

Broger, C., Institut für Biochemie und Molekularbiologie der Universität Bern, Bühlstraße 28, CH-3012 Bern

Capitanio, N., Institute of Biological Chemistry, University of Bari, Piazza G. Cesare, I-70124 Bari

Casali, E., Institute of Biological Chemistry, University of Parma, I-43100 Parma

Cavatorta, P., Department of Biophysics-GNCB, University of Parma, I-43100 Parma

Comaschi, V., Institute of Biological Chemistry, University of Parma, I-43100 Parma

Farruggia, G., Institute of Biological Chemistry, University of Parma, I-43100 Parma

Ferrari, M.B., Institute of Biological Chemistry, University of Parma, I-43100 Parma

Franzoni, L., Institute of Biological Chemistry, University of Parma, I-43100 Parma

Gesmundo, N., Institute of Biological Chemistry, University of Parma, I-43100 Parma

Johns, J., National Research Council, Division of Biological Sciences, Ottawa K1A OR6, Canada

Lüthy R., Institut für Biochemie und Molekularbiologie der Universität Bern,
Bühlstraße 28, CH-3012 Bern

Mahoney, C.W., Institut für Biochemie und Molekularbiologie der Universität
Bern, Bühlstraße 28, CH-3012 Bern

Masotti, L., Institute of Biological Chemistry, University of Parma,
I-43100 Parma

Montecucco, C., Institute of General Pathology, University of Padova,
I-35131 Padova

Müller, M., Institut für Biochemie und Molekularbiologie der Universität Bern,
Bühlstraße 28, CH-3012 Bern

Nałęcz, K.A., Nencki Institute of Experimental Biology, Polish Academy of
Sciences, PL-02-093 Warszawa

Nałęcz, M.J., Nencki Institute of Experimental Biology, Polsih Academy of
Sciences, PL-02-093 Warszawa

Nyrén, P., Department of Biochemistry, Arrhenius Laboratory, University of
Stockholm, S-106 91 Stockholm

Papa, S., Institute of Biological Chemistry, University of Bari, Piazza G.
Cesare, I-70124 Bari

Sartor, G., Institute of Biological Chemistry, University of Parma,
I-43100 Parma

Spisni, A., Institute of Biological Chemistry, University of Parma,
I-43100 Parma

Szabo, A.G., National Research Council, Division of Biological Sciences,
Ottawa K1A OR6, Canada

Szewczyk, A., Nencki Institute of Experimental Biology, Polish Academy of
Sciences, PL-02-093 Warszawa

Vignais, P.V., Laboratoire de Biochimie, Centre d'Etudes Nucléaires 85 X,
F-38041 Grenoble Cedex

Von Berger, J., Institute of Biological Chemistry, University of Parma,
I-43100 Parma

Wojtczak, L., Nencki Institute of Experimental Biology, Polish Academy of
Sciences, PL-02-093 Warszawa

I. Membrane Proteins: Function, Purification, and Characterization

Purification of Cytochrome c Reductase and Oxidase by Affinity Chromatography

REINHARD BOLLI, CLEMENS BROGER and ANGELO AZZI

I. Introduction

Cytochrome c reductase and oxidase have traditionally been purified from a large number of species and tissues by techniques based on detergent solubilization of these membrane enzymes and subsequent fractionation by ammonium sulfate precipitation [Nelson (1978), Yu et al. (1979); for a review on the enzymes see Azzi et al. (1985) and Rich (1984)]. These procedures are useful only for large-scale preparations and they have to be adapted for the isolation of the enzymes from different sources.

In the present experiment a more recently developed method is described. It makes use of the specific interaction between cytochrome c and the two enzymes which act as electron acceptor and electron donor for cytochrome c, respectively. The use of cytochrome c as an affinity ligand for the purification of the two enzymes was proposed already in 1960 and subsequently developed further (Weiss et al. 1978). It was known that cytochrome c contains many lysine residues, through which the protein could be attached to a cyanogen bromide-activated Sepharose 4B gel. Later, however, it was found that these lysine residues are located mainly on one side of the cytochrome c molecule ("front side") and that they are important for the interaction between this protein and its electron donor and acceptor (Margoliash and Bosshard 1983). This could have been the reason why such an affinity chromatography procedure worked only in some cases.

Here, an affinity resin is used, where cytochrome c is attached to the solid support via its "back side". Cytochrome c from *Saccharomyces cerevisiae* contains a cysteine residue near its C-terminus, which is located on the rear side of the molecule. The cysteine can be linked to Activated Thiol Sepharose 4B (Glutathion attached to cyanogen bromide activated Sepharose 4B) via a disulfide bridge (Fig. 1), thus leaving the front side of the protein exposed for the interaction with its electron donor and acceptor (Azzi et al. 1982).

Membrane Proteins, ed. by Azzi
©Springer-Verlag Berlin Heidelberg 1986

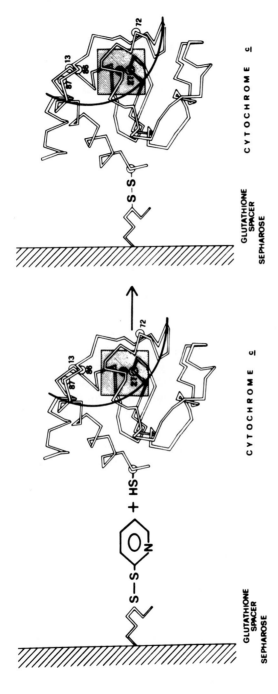

Fig. 1. Cytochrome c from *Saccharomyces cerevisiae* linked to activated thiol Sepharose 4B via a disulfide bridge

Since the surface domain on cytochrome c, which interacts with cytochrome c reductase and oxidase, is almost the same (Capaldi et al. 1982) both enzymes will bind to the affinity resin although with different affinity. Further, since the interaction is mainly electrostatic, the enzymes can be eluted from the resin by increasing the ionic strength of the buffer.

II. Equipment and Solutions

A. Equipment

— Refrigerated high speed centrifuge or ultracentrifuge
   recording spectrophotometer
— Fraction collector
— Pipettes
— Sintered glass filter (G3)
— Small columns for chromatography
— Peristaltic pump

B. Chemicals and Solutions

— Beef heart mitochondria
— Activated Thiol Sepharose 4B (Pharmacia or prepared in the laboratory from Sepharose 4B and glutathion)
— Sephadex G-25 (Pharmacia)
— Buffer: 40 mM Tris/HCl, 1 mM EDTA, pH 7.4
— Acetate buffer: 50 mM Na-acetate, 1.5 mM 2-mercaptoethanol, pH 4.5
— Cytochrome c (yeast) (Sigma)
— Cytochrome c (horse heart) (Sigma)
— Triton X-100, 20% (w/v) in water
— Tween 80, 20% (w/v) in water
— 1 M NaCl
— Sodium dithionite, solid
— 10 mM KCN
— Antimycin A (Sigma), 1mg ml^{-1} in ethanol
— Sodium borohydride, solid
— Quinole: DBH (reduced form), 10 mM in ethanol or Q238 (Aldrich)
   (DBH: 2,3-dimethoxy-5-methyl-6-decyl-1,4-benzoquinone)
   (Q238: 2-methyl-3-undecyl-1,4-naphtoquinone)

III. Experimental Procedures

A. Preparation of the Affinity Column

2 ml of swollen Activated Thiol Sepharose 4B are washed with 100 ml buffer on a sintered glass filter and incubated over night in the same buffer with 6 mg yeast cytochrome c (3 mg ml^{-1} of swollen gel). The total volume should be about 4 ml. In order to be sure that the cysteine residue of yeast cytochrome c is not blocked, it is advised to reduce the cytochrome c prior to the coupling to the gel: about 10 mg of cytochrome c is dissolved in 0.5 ml of buffer and reduced with 0.3 M 2-mercaptoethanol. After 10 min of incubation, excess reagent is removed by passing the solution through a short Sephadex G-25 column (15 x 1 cm) and collecting only the red portion of the eluate. The next morning, the affinity gel is washed free of noncovalently bound cytochrome c with 100 ml of 1 M NaCl. Activated thiol groups of the resin which did not react with cytochrome c are de-activated by washing the gel three times with 10 ml of acetate buffer containing the SH-group reagent 2-mercaptoethanol. About 60% of the added cytochrome c is bound to the gel. The gel may be stored in 1 M NaCl at 4°C for some weeks. Before use, the gel is washed with 100 ml of buffer, filled into a small column and equilibrated with 50 ml of buffer containing 1% Triton X-100.

B. Purification of Cytochrome c Oxidase and Reductase

Mitochondria (about 40 mg of protein) are diluted with the buffer to a protein concentration of 2 mg ml^{-1} and Triton X-100 is added to a final concentration of 1%. The solution is incubated for 30 min at 4°C. After centrifugation at 100,000 g for 30 min or 45 min at 50,000 g, the pellet is discarded and 10 ml of the supernatant is loaded on the column. Subsequently the column is washed with 20 ml of buffer containing 0.1% Triton X-100 at a flow rate of ca. 25 ml h^{-1}, until no heme absorbance is detected in the eluate. Fractions of 5 ml are collected.

Now the ionic strength of the elution buffer is increased to 50 mM NaCl. Cytochrome c oxidase is eluted next and four fractions of 5 ml are collected. Upon further increasing the ionic strength to 150 mM NaCl, cytochrome c reductase is eluted from the column. Again four fractions of 5 ml are collected. Spectra of all the fractions are recorded and the cytochrome content is determined as described under "measurements".

After washing extensively with lM NaCl containing 1% Triton X-100, the column can be reused although the binding capacity will be slightly reduced.

C. Measurements

1. Spectra

Difference spectra (reduced minus oxidized) of the Triton extract and of the fractions from the column are recorded in the range of 400 to 650 nm. The reduced sample is obtained by dissolving few grains of Na-dithionite directly in the cuvette. Before recording, the sample is incubated for 2 min. The cytochrome b, c_1, and aa_3 are calculated according to the law of Beer-Lambert using the extinction coefficients indicated in Table 1.

Table 1. Extinction coefficients

Cytochrome		Wavelenth	Extinction coefficient (mM)
Oxidase	aa_3	605–630 nm	27
Reductase	b	562–575 nm	25.6
Reductase	c_1	554–540 nm	20
Cytochrome	c	550–540 nm	19
		550 nm	21

2. Cytochrome c Oxidase Activity

The substrate of cytochrome c oxidase, reduced cytochrome c (ferrocytochrome c), is obtained by adding a few grains of sodium dithionite to a concentrated solution of horse heart cytochrome c in buffer. The dithionite is removed subsequently by passing the solution through a short column (15 x 1 cm) of Sephadex G-25 equilibrated in buffer. The red cytochrome c fraction of the eluate is collected and the heme c concentration is determined spectrophotometrically.

The activity of cytochrome c oxidase is measured by adding small aliquots of the fractions (10 to 50 μl) from the affinity column to a cuvette filled with 1 ml of buffer containing 0.5% Tween 80 and 5 μM ferrocytochrome c. The absorbance decrease is measured at 550 minus 540 nm or at 550 nm only, depending on the spectrophotometer available. The molecular activity is calculated as mol of cytochrome c oxidized per mol of hema aa_3 per second (turnover number).

3. Cytochrome c Reductase Activity

Reduction of Quinone (Q238). 5—10 mg of quinone are dissolved in a small
volume of a 1:1 mixture of ethanol and DMSO. Reduction is performed by
adding small portions of sodium-borohydride until the solution stays brown.
Not reacted reagent is destroyed by adding dropwise 1 M HCl, until the gas
evolution ceases. The concentration is determined from the weight of the
originally dissolved quinone and the final volume of the solution.

  Activity. A cuvette is filled with 1 ml of buffer containing 0.5% Tween
80, 5 μM oxidized horse heart cytochrome c, 0.1 mM KCN and 10 μM quinole. The rate of nonenzymatic reduction of cytochrome c is recorded as an
absorbance increase at 550 minus 540 nm (or 550 nm). The reaction is then
started by adding aliquots of the fractions from the affinity column and
after some time 2 μl of antimycin A is added. Only the antimycin-sensitive
rate of cytochrome c reduction is measured. The cytochrome c reductase
activity is given as mol of cytochrome c reduced per mol of heme b per second (turnover number).

IV. Results, Interpretation and Comments

Purification. The concentration of heme aa_3, b, and c_1 in the Triton extract
is about 0.65, 0.55, and 0.26 μM, respectively. After loading and washing of
the affinity column, 70% of cytochrome b and 55% of cytochrome oxidase is
bound, which corresponds to a binding capacity of the affinity gel of 3.5—4
nmol enzyme per ml of packed resin (2 nmol of bc_1 and 1.7 nmol of heme
aa_3). Recovery after chromatography for both enzymes is about 80%. The
cross-contamination in the eluted fractions is normally not more than 10%.
The separation, however, can be improved to zero cross-contamination by
applying a linear salt gradient between 0 and 175 mM NaCl (50 ml volume
total) instead of the described elution in steps.

  Enzyme Activity. A molecular activity of 100 mol mol^{-1} s^{-1} for cytochrome c oxidase and 10 mol mol^{-1} s^{-1} for the purified bc_1-complex using Q238
as substrate can be expected.

  The results may be best represented in a table as proposed in Table 2.

  Enzyme Purity can be further analyzed by a protein determination in the
fractions according o Lowry et al. (1951) and calculation of the heme to
protein ratio. Pure bc_1-complex and cytochrome c oxidase isolated by tra-
ditional procedures contain about 6.5—7 nmol heme b mg^{-1} protein and 5
nmol heme aa_3 mg^{-1}, respectively. This purity can also be achieved by the
method described here. Polypeptide analysis by polyacrylamide gel electro-

Table 2. Separation of bc_1-complex and cytochrome c oxidase by affinity chromatography: results

Sample	Cyt. b			Cyt. aa_3		
	nmol ml^{-1}	nmol	%	nmol ml^{-1}	nmol	%
Triton extract (loaded)	0.56	5.6	100	0.62	6.2	100
Pass through:						
Fr. 1	0.12	0.6		0.06	0.3	
Fr. 2	0.18	0.9		0.2	1.0	
Wash:						
Fr. 1	0.04	0.2		0.18	0.9	
Fr. 2	0.02	0.1		0.1	0.5	
Fr. 3	–	–		0.05	0.25	
Fr. 4	–	–		0.02	0.1	
Total not bound:	–	1.8	32	–	3.05	50
bound	–	3.8	=100	–	3.15	=100
Elution 50 mM NaCl						
Fr. 1	0.04	0.2		0.4	2.0	
Fr. 2	0.01	0.05		0.10	0.5	
Fr. 3	–	–		0.01	0.05	
Fr. 4	–	–		–	–	
Total 50 mM NaCl	–	0.25	6.5	–	2.5	81
Elution 150 mM NaCl						
Fr. 1	0.6	3.0		0.06	0.3	
Fr. 2	0.05	0.25		–	–	
Fr. 3	0.01	0.05		–	–	
Fr. 4	–	–		–	–	
Total 150 mM NaCl	–	3.3	87	–	0.3	9.5
Total eluted	–	3.55	93.5	–	2.8	89
Loss	–	0.25	6.5	–	0.35	11

phoresis should show not more than 13 main bands for cytochrome c oxidase (Kadenbach et al. 1983) and 8 to 10 for the bc_1-complex (Nałęcz et al. 1985).

References

Azzi A, Bill K, Broger C (1982) Proc Natl Acad Sci USA 79:2447–2450

Azzi A, Bill K, Bolli R, Casey RP, Nałęcz KA, O'Shea P (1985) In: Benga G (ed) Structure and properties of cell membranes, vol II. CRC Press, Boca Raton, Florida, pp 105–138

Capaldi RA, Darley-Usmar V, Fuller S, Millett F (1982) FEBS Lett 138:1–7

Kadenbach B, Jarausch J, Hartman R, Merle P (1983) Anal Biochem 129:517–521

Lowry OH, Rosenbrough NY, Farr AL, Randall RJ (1951) J Biol Chem 193:265–275

Margoliash E, Bosshard HR (1983) TIBS Sept: 316–320

Nałęcz MJ, Bolli R, Azzi A (1985) Arch Biochem Biophys 236:619–628

Nelson BD, Gellerfors P (1978) Methods Enzymol 53:80–91

Rich PR (1984) Biochim Biophys Acta 768:53–79

Weiss H, Juchs B, Ziganke B (1978) Methods Enzymol 53:98–112

Yu C, Yu L, King TE (1979) J Biol Chem 250:1383–1392

Molecular Weight Estimation of Membrane Proteins: Gel Permeation Chromatography and Sucrose Gradient Centrifugation

KATARZYNA A. NAŁĘCZ, REINHARD BOLLI and ANGELO AZZI

I. Introduction

The determination of the overall state of association of a protein requires the measurement of its molecular weight. In the case of a globular, water-soluble protein, this can be done easily and accurately by a gel filtration expe⁻ ntn, in which the permeation of the protein into the pores of a gel matrix is compared with that of proteins of known size and molecular weight. A solubilized membrane protein, however, is a detergent-protein-complex, which often contains a considerable amount of bound lipid. To estimate the molecular weight of the protein moiety in such a complex, one has first to determine the amounts of each of the components bound, and one has to know also their partial specific volumes. Gel permeation chromatography and centrifugation techniques may then be applied to determine size, molecular weight, and aggregation state of the protein complex. The obtained result has to be interpreted carefully, considering especially the asymmetry of the protein complex, the possible interaction of the detergent with the gel matrix, and the influence of other components of the medium (e.g., pH, ionic strength).

In the following experiment gel filtration and sucrose gradient centrifugation are applied to determine the molecular weight and aggregation state of cytochrome c oxidase from bovine heart mitochondria and from *Paracoccus denitrificans*. These two enzymes have similar functions, but rather different structures and subunit composition (for reviews see Azzi et al. 1985, Azzi 1980, Ludwig 1980). Although the activity of purified bovine cytochrome c oxidase is quite poor in Triton X-100, this detergent has the advantage of dispersing these proteins at very low concentration and its binding to the enzymes can easily be measured spectrophotometrically avoiding the use of radioactive substances. The lipid content of the isolated enzymes was shown to be very low (Nałęcz et al. 1985). Therefore the contribution of the bound lipid to the molecular weight of the detergent-protein complex can be neglected.

Membrane Proteins, ed. by Azzi
© Springer-Verlag Berlin Heidelberg 1986

Table 1. Standard error function (erf)

erf^{-1}	$1-K_d$		$1-K_d$		$1-K_d$		$1-K_d$
0,01	0,0112833	0,51	0,5292437	1,01	0,8468105	1,51	0,9672768
0,02	0,0225644	0,52	0,5378987	1,02	0,8508380	1,52	0,9684135
0,03	0,0338410	0,53	0,5464641	1,03	0,8547842	1,53	0,9695162
0,04	0,0451109	0,54	0,5549392	1,04	0,8586499	1,54	0,9705857
0,05	0,0563718	0,55	0,5633233	1,05	0,8624360	1,55	0,9716227
0,06	0,0676215	0,56	0,5716157	1,06	0,8661435	1,56	0,9726281
0,07	0,0788577	0,57	0,5798158	1,07	0,8697732	1,57	0,9736026
0,08	0,0900871	0,58	0,5879229	1,08	0,8733261	1,58	0,9745470
0,09	0,1012806	0,59	0,5959365	1,09	0,8768030	1,59	0,9754620
0,10	0,1124630	0,60	0,6038561	1,10	0,8802050	1,60	0,9763484
0,11	0,1236230	0,61	0,6116812	1,11	0,8835330	1,61	0,9772069
0,12	0,1347584	0,62	0,6194114	1,12	0,8867879	1,62	0,9780381
0,13	0,1458671	0,63	0,6270463	1,13	0,8899707	1,63	0,9788429
0,14	0,1569470	0,64	0,6345857	1,14	0,8930823	1,64	0,9796218
0,15	0,1679959	0,65	0,6420292	1,15	0,8961238	1,65	0,9803756
0,16	0,1790117	0,66	0,6493765	1,16	0,8990962	1,66	0,9811049
0,17	0,1899923	0,67	0,6566275	1,17	0,9020004	1,67	0,9818104
0,18	0,2009357	0,68	0,6637820	1,18	0,9048374	1,68	0,9824928
0,19	0,2118398	0,69	0,6708399	1,19	0,9076083	1,69	0,9831526
0,20	0,2227025	0,70	0,6778010	1,20	0,9103140	1,70	0,9837904
0,21	0,2335218	0,71	0,6846654	1,21	0,9129555	1,71	0,9844070
0,22	0,2442958	0,72	0,6914330	1,22	0,9155339	1,72	0,9850028
0,23	0,2550225	0,73	0,6981038	1,23	0,9180501	1,73	0,9855785
0,24	0,2657000	0,74	0,7046780	1,24	0,9205052	1,74	0,9861346
0,25	0,2763263	0,75	0,7111556	1,25	0,9229001	1,75	0,9866717
0,26	0,2868997	0,76	0,7175367	1,26	0,9252359	1,76	0,9871903
0,27	0,2974182	0,77	0,7238216	1,27	0,9275136	1,77	0,9876910
0,28	0,3078800	0,78	0,7300104	1,28	0,9297342	1,78	0,9881742
0,29	0,3182834	0,79	0,7361035	1,29	0,9318987	1,79	0,9886406
0,30	0,3286267	0,80	0,7421010	1,30	0,9340080	1,80	0,9890905
0,31	0,3389081	0,81	0,7480033	1,31	0,9360632	1,81	0,9895245
0,32	0,3491259	0,82	0,7538108	1,32	0,9380652	1,82	0,9899431
0,33	0,3592785	0,83	0,7595238	1,33	0,9400150	1,83	0,9903467
0,34	0,3693644	0,84	0,7651427	1,34	0,9419137	1,84	0,9907359
0,35	0,3793819	0,85	0,7706680	1,35	0,9437622	1,85	0,9911110
0,36	0,3893296	0,86	0,7761002	1,36	0,9455614	1,86	0,9914725
0,37	0,3992059	0,87	0,7814398	1,37	0,9473124	1,87	0,9918207
0,38	0,4090093	0,88	0,7866873	1,38	0,9490160	1,88	0,9921562
0,39	0,4187385	0,89	0,7918432	1,39	0,9506733	1,89	0,9924793
0,40	0,4283922	0,90	0,7969082	1,40	0,9522851	1,90	0,9927904
0,41	0,4379690	0,91	0,8018828	1,41	0,9538524	1,91	0,9930899
0,42	0,4474676	0,92	0,8067677	1,42	0,9553762	1,92	0,9933782
0,43	0,4568867	0,93	0,8115635	1,43	0,9568573	1,93	0,9936557
0,44	0,4662251	0,94	0,8162710	1,44	0,9582966	1,94	0,9939229
0,45	0,4754818	0,95	0,8208908	1,45	0,9596950	1,95	0,9941794

Table 1. Continuation

0,46	0,4846555	0,96	0,8254236	1,46	0,9610535	1,96	0,9944263
0,47	0,4937452	0,97	0,8298703	1,47	0,9623729	1,97	0,9946637
0,48	0,5027498	0,98	0,8342315	1,48	0,9636541	1,98	0,9948920
0,49	0,5116683	0,99	0,8385031	1,49	0,9648979	1,99	0,9951114
0,50	0,5204999	1,00	0,8427008	1,50	0,9661052	2,00	0,9953223

II. Theory and Calculation

A. Gel Filtration

Gel permeation chromatography is a convenient method to characterize the size of a detergent-protein complex in terms of its Stokes' radius. However, it does not give direct information about the molecular weight and the Stokes' radius obtained is an apparent radius, since a possible retardation of the complex or its further exclusion from the gel matrix due to its asymmetry or interaction with the gel have to be considered. Radius measurements using other techniques (e.g., light scattering) should confirm the results obtained from gel filtration experiments. A convenient way for the presentation of the chromatographic data is based on the following equation (Ackers 1967):

$$R_S = a_o + b_o \, \mathrm{erf}^{-1} \, (1-K_d), \tag{1}$$

where R_S = Stokes' radius

a_o, b_o = constants of the gel matrix used

erf^{-1} = inverse error function

K_d = partition coefficient = $(V_e-V_o)/(V_t-V_o)$

where V_e. V_o and V_t are the elution volume of the protein, the void- and total volume of the column, respectively.

The values of erf^{-1} for the measured $(1-K_d)$ can be read from Table 1 (standard error function: see also Dwight 1961), by looking for the argument of the function's value $(1-K_d)$. a_o and b_o can be estimated from the chromatography of two calibration proteins of known R_S. Figure 1 shows the calibration of the Ultrogel AcA 34 column used in this experiment. Table 2 gives the Strokes' radii of some proteins suitable for column calibration (see also Le Maire et al. 1980).

Ultrogel AcA 34 separates cytochrome c oxidase (bovine and from *P. denificans*) sufficiently well from an excess of Triton X-100 micelles. This enables the quantitation of the detergent molecules bound tightly to the protein. In the eluted fractions, which contain the enzyme, the ratio r of the absorptions at 277 nm and 422 nm is increased compared to that of the pure protein (mea-

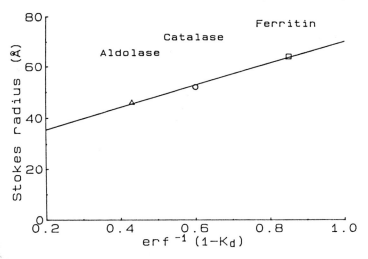

Fig. 1. Calibration of the Ultrogel AcA 34 column

Table 2. Stokes' radius of calibration proteins

Calibration protein	Molecular weight	Stokes' radius ($\overset{\circ}{A}$)
Cytochrome c	13,400	17
Ribonuclease	13,700	17.5
Myoglobin	16,900	19
Ovalbumin	43,000	28
Alkaline phosphatase	86,000	33
Transferrin	81,000	36
Aldolase	158,000	46
Catalase	230,000	52
Ferritin	440,000	64
Thyroglobulin	660,000	86

sured, e.g., in cholate or dodecyl maltoside, detergents which do not absorb at 277 nm). Knowing the extinction coefficients of Triton X-100 at 277 nm and of the oxidase at 422 nm, the amount of bound Triton can be calculated from the equation:

$$\text{mol TX-100/mol oxidase} = (r_{complex} - r_{protein}) \cdot \epsilon^{ox}_{422}/\epsilon^{Tx}_{277} \qquad (2)$$

$$\epsilon \, \substack{Tx \\ 277} = 1.465 \text{ mM}^{-1} \text{ cm}^{-1}$$

for the bovine enzyme:

$$\epsilon \, \substack{OX \\ 422} = 168 \text{ mM}^{-1} \text{ cm}^{-1}$$

$r_{\text{protein}} = 2.5$

for the *Paracoccus* enzyme (A_{424} is measured instead of A_{422}):

$$\epsilon \, \substack{OX \\ 424} = 163 \text{ mM}^{-1} \text{ cm}^{-1}$$

$r_{\text{protein}} = 1.8.$

B. The Partial Specific Volume

In order to calculate the molecular weight of a particle from any centrifugation analysis, one has to know its partial specific volume, which is the change
in volume of a solution due to the dissolution of the substance (protein) at
constant temperature and pressure. v is normally expressed in $\text{cm}^3 \text{ g}^{-1}$ and
is the reciprocal of the particle density. Density measurements are the best
approach to estimate v (see Nałęcz et al. 1986, Steele et al. 1978, Lee et al.
1979). The partial specific volume v* of a multicomponent complex can be
calculated by summing up the partial specific volumes v_{ijk} in the complex:

$$v^* = \Sigma \, v_{ijk} \cdot x_{ijk}. \tag{3}$$

In Table 3 the partial specific volumes of some detergents and proteins are
listed.

C. Sucrose Gradient Centrifugation and Molecular Weight

The molecular weight M* of a complex is calculated from the equation:

$$M^* = s_{20,w} \cdot RT \, / \, D_{20,w} \, (1 - v^* \, \rho_s), \tag{4}$$

where
$s_{20,w}$	= sedimentation coefficient, normalized to 20°C and water
$D_{20,w}$	= diffusion constant of the complex, normalized to 20° and water
v^*	= partial specific volume of the complex
ρ_s	= density of the solvent
R, T	= gas constant, absolute temperature.

$D_{20,w}$ can be mesured in an analytical ultracentrifuge or it may be repla-
ced by the following expression with the Stokes' radius R_S as a parameter:

$$D = kT/6\pi\eta R_S. \tag{5}$$

Table 3. Partial specific volumes

	v (cm^3 g^{-1})	Reference
Detergents		
SDS	0.87	Tanford and Reynolds (1976)
Deoxycholate	0.778	Tanford and Reynolds (1976)
Triton X-100	0.908	Tanford and Reynolds (1976)
Triton N-101	0.922	Tanford and Reynolds (1976)
Tween-20	0.869	Tanford and Reynolds (1976)
Tween-80	0.896	Tanford and Reynolds (1976)
Lubrol PX	0.958	Tanford and Reynolds (1976)
Dodecyl-maltoside	0.815	Nałęcz et al. (1986)
Proteins		
Ribonuclease	0.709	McMeekin and Marshall (1952)
Fibrinogen (human)	0.725	McMeekin and Marshall (1952)
BSA	0.734	McMeekin and Marshall (1952)
Adolase	0.74	McMeekin and Marshall (1952)
Hemoglobin	0.749	McMeekin and Marshall (1952)
Ovalbumin	0.745	McMeekin and Marshall (1952)
Cytochrome c oxidase		
Bovine 5°C	0.822	Bolli et al. (1985)
20°C	0.763	Bolli et al. (1985)
Paracoccus 20°C	0.551	Nałęcz et al. (1985)

Equation (4) then becomes:

$$M^* = s_{20,w} \cdot R_S \cdot (6\pi\eta N) / (1 - v^*\rho_s) \tag{6}$$

N = Avogadro number
η = viscosity of the solvent.

The sedimentation coefficient can be estimated from a sedimentation analysis in an analytical ultracentrifuge or, more simply, from a sucrose gradient centrifugation. The sedimentation velocity dr/dt is defined by the equation:

$$dr/dt = s(r) \, \omega^2 r \tag{7}$$

    r = distance from the rotor axis
  $s(r)$ = sedimentation coefficient
   ω = angular velocity.

In a linear sucrose gradient the distance from the rotor axis can be expressed in terms of the sucrose concentration z. The theoretical sucrose concentration at the rotor axis would be z_0.

The sedimentation coefficient of a particle at a certain sucrose concentration is a function of the viscosity and the density of the solution. The tables

Table 4. Values of time integral (I_z) for sucrose gradient centrifugation. Temperature: 4°C; theoretical sucrose concentration at the rotor axis (z_0): -10%

Sucrose concentration (z) (%)	Integral I_z	
	$\rho = 1.2$ g cm^{-3}	$\rho = 1.4$ g cm^{-3}
0	0.0000	0.0000
2	0.2906	0.2854
4	0.5619	0.5461
6	0.8228	0.7913
8	1.0800	1.0269
10	1.3389	1.2578
12	1.6043	1.4876
14	1.8810	1.7196
16	2.1744	1.9569
18	2.4905	2.2025
20	2.8365	2.4598
22	3.2216	2.7322
24	3.6580	3.0239
26	4.1620	3.3396
28	4.7565	3.6849
30	5.4754	4.0670

of McEwen (Table 4) give the integrals (I_z) for some centrifugation temperatures and particle densities for any sucrose concentration. They are normal-
ized to the density of water and 20 $^\circ$C. From the position $z(2)$ of the particle in the sucrose gradient after a certain time t of centrifugation, one can
then calculate the sedimentation coefficient according to

$$s_{20, w} = (I_{z(2)} - I_{z(1)}) / \omega^2 t. \tag{8}$$

$z(1)$ is normally the starting sucrose concentration of the gradient.

III. Equipment and Solutions

A. Gel Filtration

- Sintered glass filter
- Column, 50 x 1 cm
- Pump 5 ml h^{-1}
- Fraction collector
- UV/VIS-spectrophotometer

- Ultrogel AcA 34 (LKB), 40 ml
- Buffer A: 10 mM Tris-HCl pH 7.4; 50 mM KCl; 0.05% Triton X-100
- 10% Triton X-100
- Cytochrome c oxidase, bovine and from *Paracoccus,* ca. 220 μM heme aa3 in 50 mM Na-phosphate pH 7.4, frozen
- High molecular weight gel filtration calibration kit (Pharmacia)

B. Sucrose Gradient Centrifugation

- Ultracentrifuge
- Swing-out rotor 60, 6 x 4-ml tubes
- Gradient mixer (50 ml)
- Pump, adjustable 0.5–10 ml min^{-1}
- Gradient-forming device (to form 6 gradients in parallel)
- Gradient-harvesting device with UV-monitor and recorder
- Syringe 200 μl
- Buffer B: 10 mM Tris-HCl pH 7.4; 50 mM KCl; 0.1% Triton X-100; 5% sucrose.
- Buffer C: like buffer B, but 20% sucrose
- Cytochrome c oxidase preparations as described under III. A.

IV. Experimental Procedures

A. Gel Filtration

Packing of the Column. 40 ml of wet Ultrogel AcA 34 is first washed with 250 ml buffer A on a sintered glass filter. The gel is then suspended in 100 ml of the same buffer and cooled to 4°C. The slurry is poured into the column equipped with a reservoir container and the gel is allowed to sediment over night at a flow rate of 5 mlh^{-1}. With the same rate the column is further equilibrated with buffer A for at least 24 h.

Calibration is performed in two runs. First, a sample (150 μl) of a mixture of blue dextran, ferritin, aldolase, and ferricyanide in buffer A is loaded and in a second run blue dextran, catalase, and ferricyanide are chromatographed. The K_d and erf^{-1} $(1-K_d)$ of the calibration proteins are calculated from the measured void volume V_o (elution volume of blue dextran), the total volume V_t (ferricyanide) and the elution volumes V_e of the proteins according to Eq. (1). A calibration curve is plotted as shown in Fig. 1.

Chromatography of Cytochrome c Oxidase. 10 μl of freshly thawed cytochrome c oxidase is given to 100 μl of buffer A, which was adjusted before with 7.5 μl of 10% Triton X-100 to a final concentration of 0.75%, and incubated for 2 h in the dark at 4°C. The oxidase is then loaded on the equilibrated Ultrogel column. Elution is performed with buffer A at a flow rate of 5 ml h^{-1}. Fractions of about 0.8 ml (10 min/fraction) are collected with the fraction collector. In all fractions the absorbance at 277, 422 (or 424 for *Paracoccus* oxidase) and 480 nm is measured against buffer A. An elution profile is drawn, erf^{-1}(1-K$_d$) is calculated for the enzyme and the Triton X-100 micelles and the Stokes' radii are read from the calibration curve. In the fractions showing heme absorption the amount of the bound detergent is calculated according to Eq. (2).

B. Sucrose Gradient Centrifugation

A linear sucrose concentration gradient is formed using a device, schematically shown in Fig. 2. 13 ml of buffer B is placed into the mixing chamber and 13 ml of buffer C in the reservoir. Six gradients of 4 ml each are then formed in parallel directly in the centrifugation tubes at a pump speed of 0.5 ml min^{-1}. Twice 5 μl of each cytochrome c oxidase (bovine and *P. denitrificans*) are diluted with 100 μl of buffer A, which has been adjusted with 10% Triton X-100 to a final concentration of 0.55%, and incubated for 30 min in ice. Two samples of each cytochrome oxidase preparation are carefully layered with a syringe onto the top of four gradients. The other two gradients are loaded with 110 μl of water. The gradients are carefully transferred to the precooled rotor buckets, which are then attached to the rotor. Centrifugation is performed at 55,000 rpm at 5°C for 4 h 30 min.

After centrifugation, the tubes are removed from the rotor and the visible bands of the cytochrome c oxidase, as well as the beginning of the gradients, are marked on the tubes. The content of the tubes are discarded, the tubes rinsed with water and then well dried. From a calibrated pipet, water is poured to the tubes up to the level of the marks and the volume is noted. Assuming the sucrose concentration as a function of the volume being linear from 18.5% at the bottom to 4.5% at the top of the gradient (diffusion of sucrose during centrifugation), the sucrose concentration (z) at the marked positions can be calculated from the measured volume. In order to analyze more exactly the positions of the visible proteins and to detect non-colored proteins, one may displace the gradient continuously by pumping buffer C to the bottom of the tube. The effluent leaving the tube at the top through a tubing is then monitored at 280 nm by a UV detection unit (UVICORD, LKB) connected to a recorder. Another method would be to puncture the tube at the bottom with a needle and let the effluent pass through the bottom. On the recorded profiles, bottom and top of the gradients are signed, the position of the protein is estimated in terms of the volume from the bottom and the corresponding sucrose concentration is calculated as described above. The sedimentation coefficients ($s_{20,w}$) are calculated according to Eq. (8) using Table 4 with the following parameters:

— $z_0 = -10\%$
— temperature: $5\degree C$
— particle density = $1/v^* = 1.2 \text{ g cm}^{-3}$ (bovine cyt. c oxidase)
                               = 1.4 g cm^{-3} (*P. denitrificans*)
— $\omega^2 = 3.28 \times 10^7 \text{ s}^{-2}$

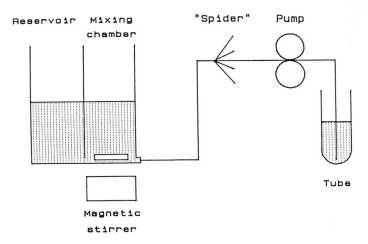

Fig. 2. Sucrose gradient-forming equipment. The solution with the higher sucrose concentration is placed into the reservoir, the one with the lower concentration into the mixing chamber. Only the solution in the mixing chamber is stirred. To obtain identical gradients it is advised to use a multi-channel peristaltic pump, which can be placed after the "spider"

C. Calculation of the Molecular Weight

Equation (6) is used to calculate the molecular weight of the Triton X-100-protein complex. R_S is taken from the gel filtration analysis, the product of the constants $6 \pi \eta N$ is 113.1×10^{21} and the density of the medium can be approximated with 1.0 g cm^{-3}. The partial specific volume of the complex (v^*) has to be calculated from Eq. (3). v of the proteins are taken from Table 3 and the weight fraction of Triton is calculated from the binding mea-

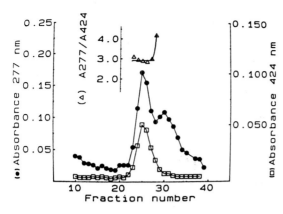

Fig. 3. Gel filtration of cytochrome c oxidase from *Paracoccus denitrificans* on Ultrogel AcA 34. 1.6 nmol of heme aa$_3$ in 120 μl was incubated in 0.3% Triton X-100, loaded and eluted with buffer A. Fractions of 0.8 ml were collected. The absorbance at 424 nm (▢) and at 277 nm (●) was measured and in the peak fractions the A$_{277}$/A$_{424}$ ratio (△) calculated

surements by gel filtration, assuming an average molecular weight of 628 for Triton X-100. The molecular weight of the protein moiety can be obtained by subtracting the molecular weight fraction of the bound detergent. The aggregation state of the enzyme has to be worked out considering a minimal molecular weight of 200,000 for the bovine and 82,000 for the *Paracoccus* oxidase monomer, known from gel electrophoresis and amino acid sequencing data.

V. Results and Comments

Experimentally determined and calculated parameters necessary to estimate the molecular weights of bovine and *Paracoccus* cytochrome c oxidase are summarized in Table 5. As an example of a gel filtration experiment, the elution profile for the *Paracoccus* enzyme is shown in Fig. 3. The central fractions of the eluted protein represented by its heme absorption at 424 nm give a constant A$_{277}$/A$_{424}$-ratio (2.85) reflecting a homogeneous population of complexes with a constant amount of bound Triton X-100 (111 mol mol^{-1} heme aa$_3$). This value should not change if the protein/detergent ratio is changed during preincubation. The amount of detergent added should be sufficient to saturate the protein, but not in such an excess that the free detergent micelles are no longer well separated from the protein complex.

The partial specific volumes of cytochrome c oxidase given in Table 3 are determined by density measurements. Knowing the amino acid composition of a protein, one may alternatively estimate v by summing up the partial specific volumes of the amino acids according to their weight fractions and fre-

Table 5. Molecular weight analysis of cytochrome c oxidase from bovine heart mitochondria and *Paracoccus denitrificans*

Parameter	Value	
	Paracoccus	Bovine
K_d	0.36	0.25
$erf^{-1}(1-K_d)$	0.65	0.825
R_S (Å)	54	63
$r_{complex}$	2.85	4.15
Bound Triton X-100 (mol/mol)	111	190
Molecular weight fraction of the bound detergent per heme aa_3	70,000	120,000
Partial specific volume of the complex (cm^3/g)	0.717	0.817
Sedimentation coefficient $s_{20,w}$ (S)	7.2	15.5
Molecular weight M*	155,000	600,000
Aggregation state	Monomer	Dimer
Molecular weight of the protein moiety	85,000	360,000

quency in the protein (see Lee et al. 1979). The values calculated in this way and those determined experimentally by density measurements are similar for globular proteins, but may differ considerably in the case of asymmetric, hydrophobic proteins.

In order to calculate the molecular weight of the protein moiety, having determined the molecular weight of the protein-detergent complex, one has to subtract the contribution of the bound detergent. This value, however, depends on the aggregation state of the protein. The easiest way to estimate the aggregation state is to calculate the molecular weights of the monomeric, dimeric, trimeric, etc. protein-detergent complexes considering the minimal molecular weight known from SDS polyacrylamide gel electrophoresis as the value for the protein moiety fraction of the monomer and taking into account the corresponding amount of bound detergent. One of these calculated molecular weights should then correspond to the one experimentally determined. In the case of the *Paracoccus* oxidase this is clearly the monomer, and for the bovine enzyme the best fit is the dimer. In the latter case the molecular weight has therefore been slightly underestimated.

References

Ackers GK (1967) J Biol Chem 242:3227–3228
Azzi A (1980) Biochim Biophys Acta 594:231–252
Azzi A, Bill K, Bolli R, Casey RP, Nałęcz KA, O'Shea P (1985) In: Benga G (ed) Structure and properties of cell membranes, vol. II. CRC Press, Boca Raton, Florida, pp 105–138
Bolli R, Nałęcz KA, Azzi A (1985) Arch Biochim Biophys 240:102–116
Dwight HB (1961) Mathematical tables. Dover New York, pp 140–143
Lee JC, Gekko K, Timasheff SN (1979) Methods Enzymol 61:26–57
Le Maire M, Rivas E, Moeller JV (1980) Anal Biochem 106:12–21
Ludwig B (1980) Biochim Biophys Acta 594:177–189
McEwen CR (1967) Anal Biochim 20:114–149
McMeekin TL, Marshall K (1952) Science 116:142
Nałęcz KA, Bolli R, Ludwig B, Azzi A (1985) Biochim Biophys Acta 808:259–272
Nałęcz KA, Bolli R, Azzi A (1986) Methods Enzymol 126:45–64
Steele JCH, Tanford C, Reynolds JA (1978) Methods Enzymol 48:11–29
Tanford C, Reynolds JA (1976) Biochim Biophys Acta 457:133–170

The Problem of Light Scattering in the Absorbance and Fluorescence Studies of Proteins in Membranes

P. CAVATORTA, E. CASALI and G. SARTOR

I. Introduction

Light scattering by suspension of biological particles has proven to be a convenient indicator of some physiological processes. Changes in rates of phosphorylation, electron transport, etc., and changes in the osmolarity of the medium can modify the scattering properties of mitochondria. Presumably the changes in scattering properties reflect changes in particle conformation (Latimer et al. 1968).

For this reason, the light scattering technique may be a very useful although complex method to evaluate the sizes and changes in shape of biological samples.

On the other hand, light scattering represents a very serious handicap in the achievement of reproducible results using other spectroscopic techniques, such as absorption, fluorescence, circular dichroism etc.

As the scatter intensity is directly related to the concentration and dimensions of the scattering particles, it is clear that the problem is not particularly relevant for dilute solutions of small molecules but it becomes important for cells, membranes, or large macromolecules like nucleic acids, polypeptides and proteins.

Furthermore, as it is well established that the scatter intensity is inversely related to the wavelength, the scattering artifacts become particularly effective with proteins and nucleic acids whose intrinsic chromophores absorb in the ultraviolet region.

The artifacts caused by light scattering in obtaining circular dicroism spectra of membrane proteins have been elucidated by Urry et al. (1970) and recently by Wallace and Mao (1984).

The aim of the present experiments is limited to studying the distortions caused by light scattering in absorption and fluorescence experiments on membranes and to indicate some empirical methods to correct them.

Membrane Proteins, ed. by Azzi

Absorption Spectroscopy

Absorption measurements on membrane proteins are distorted by scattering arising from the particulate nature of the membranes. In circular dichroism studies, in fluorescence quantum yield determination etc. it is crucial to know the exact concentration or the exact optical density of the absorbing species, therefore it is fundamental to obtain good absorption spectra. If we have a suitable blank, exactly equal to the membrane's matrix, we can use it as a reference, and by subtraction from the protein lipid complex spectrum we will obtain the absorption spectrum of the protein alone. Unfortunately, very often this is not possible because the incorporation of proteins in model membranes causes changes in the shape and dimensions of the particles.

For example, the incorporation of gramicidin A' in lysolecithin results in a change from the micellar aggregation of the lipid to a bilayer structure with a dramatic variation in dimension and shape of the particles (Spisni et al. 1983). In some fortunate cases, it is possible to correct the scattering artifacts using a simple semi-empirical method. It is well established that in a medium containing spherical scatterers (no absorbers), the spectral transmittance over a pathlength of 1 cm is $T = e^{-\gamma}$, where γ is the scattering coefficient. For small particles with respect to the wavelength, γ is proportional to λ^{-4} (Rayleigh scattering), while for larger particles, γ is proportional to λ^{-K} (Mie scattering) (Hudson 1969).

The Mie relation may be written as:

$$\log A = K \log \lambda + C,$$

where A is the absorbance simulated by the scattering, and C and K are constants. Taking into consideration that the intrinsic chromophores of the proteins, in the absence of any prostetic group, do not absorb light above 310 nm, any absorbance detected at longer wavelengths is due to the scattering.

It is possible to plot several absorbance values vs. the wavelength, and with a program of linear least-squares analysis, to calculate the best value of K and C for the system of interest.

Furthermore it is possible, by extrapolation to the shorter wavelengths, to find the scattering values, and by subtraction, to reconstruct the true absorption spectrum of the protein. The question to be answered now is if the Mie equation is a good approximation for the scattering of natural and model membranes.

In Fig. 1 we show the experimental and simulated plots of the scattering of vesicles of dimirystoylphosphatidylcholine (DMPC). The values of the statistical parameters guarantee the applicability of the Mie relation in this case.

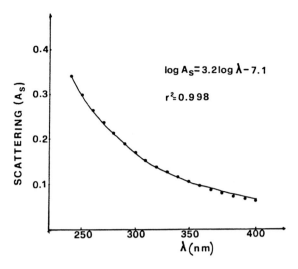

Fig. 1. Experimental (*solid line*) and simulated (●) plots of the scattering of DMPC vesicles. The simulated points have been calculated by the formula reported in the figure, that is the best Mie relation fitting the experimental data

II. Materials and Instruments

— N-acetyl-tryptophanamide (NATA)
— Tris buffer 20 mM pH 7
— Dimirystoylphosphatidylcholine (DMPC)
— UV-Vis spectrophotometer
— Sonicator
— Bench centrifuge

III. Experimental Procedure

The experiment consists in the simulation of an absorption spectrum of proteins in membranes. The absorption of a protein is simulated using N-acetyl-tryptophanamide (NATA) that has the same spectroscopic properties of tryptophan in proteins. At first the absorption spectrum of NATA in Tris buffer 20 mM pH 7 is detected with a spectrophotometer (Fig. 2). The concentration of NATA is not crucial; an absorbance of nearly 0.3 at 280 nm will be good for the spectrophotometric detection.

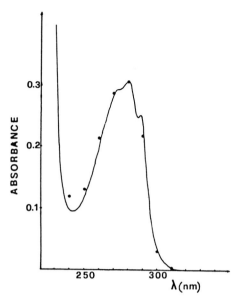

Fig. 2. Experimental (*solid line*) and simulated (●) absorption spectra of NATA. The simulated points have been obtained from Fig. 3, subtracting the scattering contribution due to DMPC vesicles

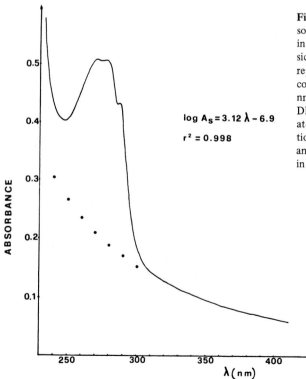

$$\log A_S = 3.12\,\lambda - 6.9$$

$$r^2 = 0.998$$

Fig. 3. Experimental absorption spectra of NATA in presence of DMPC vesicles (*solid line*). *Points* represent the scattering contribution between 240 nm and 300 nm due to the DMPC vesicles as evaluated from the best Mie relation obtained from 310 nm and 400 nm, also reported in the figure

The second step is the preparation of a vesicle suspension of DMPC. The DMPC powder is suspended in Tris 20 mM pH 7 (nearly 5 mg ml^{-1}) and sonicated under nitrogen atmosphere at 35°C for 30 min. The sonication is done in 3—min steps with intervals of 1 min. The suspension is then centrifuged for 30 min at 1500 g in order to sediment any large aggregates.

The next step is to prepare two solutions of NATA at the same concentration as in step 1 but in presence of DMPC vesicles. The stock solution of DMPC is diluted to give an "absorbance" due to the scattering near to 0.3 at 280 nm.

The resulting absorption spectrum will be similar to the absorption spectrum of a protein in a membrane (Fig. 3). Now the scattering can be corrected using the Mie relation between 310 and 400 nm and extrapolating to 240 nm. The result that must be achieved, is to obtain, by subtraction, the same absorption spectrum of NATA in solution (Fig. 2, filled circles). The value of the correlation coefficient and the closeness of the two spectra of NATA guarantee the applicability of the Mie relation in this case.

Fluorescence Spectroscopy

The scattering light may influence the fluorescence spectra in several ways. In particular the scattering can be a serious problem both in static and dynamic fluorescence measurements, and especially in fluorescence anisotropy technique (Lakowicz 1983). Since the static fluorescence anisotropy has been extensively used in protein conformation studies because of its simplicity, we will limit the present experiment to this case. In fluorescence anisotropy experiments, normally the sample is excited by vertically polarized light and the fluorescence is detected by a polarizer placed alternatively parallel (V) and perpendicular (H) with respect to the direction of the incident light.

The anisotropy is then defined as:

$$r = (I_{VV} - I_{VH}G) / (I_{VV} + 2I_{VH}G),$$

where G is a correction factor.

If the fluorescence intensities contain some scattered light, the r value can increase dramatically, because the scattering is very highly polarized. As in the absorption case, if we have a suitable blank, it is possible to correct for the scattering contribution by simply modifying the anisotropy formula as follows:

$$r = \frac{(I_{VV} - I^s_{VV}) - (I_{VH} - I^s_{VH})G}{(I_{VV} - I^s_{VV}) + 2(I_{VH} - I^s_{VH})G},$$

where I^S_{VV} and I^S_{VH} are respectively the vertically and horizontally polarized scattering intensities.

If the blank is not available, probably the only way to solve the problem is to prepare a reference by using some scattering substances (like glycogen or DMPC vesicles) that have the same scattering behavior and the same "absorbance" at the excitation wavelength of the sample, as determined by the Mie relation. Obviously the same considerations are valid also in time-resolved fluorescence experiments. The scattering may also influence the fluorescence anisotropy measurements in a way such as to decrease the true fluorescence anisotropy. As pointed out by Teale (1969), when working with turbid samples, the vertically polarized excitation light and the fluorescence light can undergo several reflections both before absorption and before reaching the detector. This fact results in a severe depolarization of the fluorescence and the observed anisotropy (r') is expected to decrease linearly with the optical density (OD) due to turbidity:

$$r' = r - r\ 2.303\ OD.$$

In this equation r is the anisotropy, which is not affected by scattering.

Recent studies have confirmed the predicted linear dependence of r' on the turbidity (Lentz et al. 1979). However, the actual proportionality constant depends upon the individual sample under investigation. Therefore, it is advisable to determine the effect of turbidity for each sample. This is generally accomplished by dilution.

IV. Materials and Instruments

- Tris buffer 20 mM pH 7
- Gramicidin A, (GA) (powder)
- Dimirystoylphosphatidylcholine (DMPC)
- 1,6-diphenylhexatriene (DPH) in Tetrahydrofuran (THF) (2 mM)
- Spectrophotofluorimeter with two polarizers
- Spectrophotometer
- Sonicator
- Bench centrifuge

V. Experimental Procedure

The aim of this experiment is to verify the Teale relation in lipid model membranes and in model membranes containing proteins. The protein effect will

be simulated by the pentadecapeptide gramicidin A, (GA) that incorporates
into lipids (Spisni et al. 1983).

The vesicles are prepared as in the first experiment by sonication at 35°C
for 30 min of a suspension of DMPC in Tris buffer, with and without GA at a
molar ratio of 10 to 1 (DMPC/GA). After the sonication, the suspension is
centrifuged for 30 min at 1500 rpm in a bench centrifuge in order to sediment the large aggregates formed. From the supernatant, several samples characterized by varying turbidity are obtained by dilution, and the optical densities are determined with a spectrophotometer at 360 nm (excitation wavelength).

A very small amount of a concentrated sample of 1,6-diphenylhexatriene
(DPH) in tetrahydrofuran (THF) (not more than 5 μl) is then added to each
sample to a final dye-to-lipid ratio smaller than 1:300. The samples are left
in the dark for 30 min in order to obtain the complete incorporation of DPH
in the lipid matrix.

The static fluorescence anisotropy values are obtained by a spectrophotofluorimeter equipped with two polarizers, and the values are plotted vs. the
optical density. By a simple linear least-squares analysis program, it is now
possible to evaluate the best linear fit for the experimental results and obtain
the proportionality constants as well as the true anisotropy values for the
two systems under study. In Fig. 4 the anisotropies of DPH fluorescence vs
optical density for three samples of DMPC-GA vesicles at 18°C are shown.

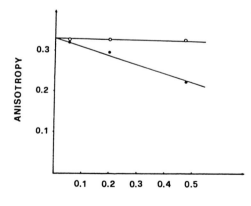

Fig. 4. Fluorescence anisotropy values
of DPH in DMPC vesicles at different
values of "absorbance (scattering) at
360 nm. (o) DMPC alone; (●) DMPC
+ Gramicidin A

References

Hudson RD (1969) Infrared system engineering. Wiley, New York
Lakowicz JR (1983) Principles of fluorescence spectroscopy. Plenum Press, London
    New York
Latimer P, Moore DM, Briant FD (1968) J Theoret Biol 21: 348−367
Lentz BR, Moore BM, Barrow DA (1979) Biophys J 25:489−494
Spisni A, Pasquali-Ronchetti I, Casali E, Lindner L, Cavatorta P, Masotti L, Urry DW
    (1983) Biochim Biophys Acta 732:58−68
Teale FWJ (1969) Photochem Photobiol 10:363−374
Urry DW, Hinners TA, Masotti L (1970) Arch Biochem Biophys 137:214−221
Wallace BA, Mao D (1984) Anal Biochem 142:317−328

Reproducible Preparation of Phosphatidylserine Vesicles for Fluorescence Studies of Protein Incorporation into Lipids

JANET JOHNS and ARTHUR G. SZABO

I. Introduction

The structure and function of proteins in lipid systems is of considerable interest. Often, in order to elucidate the details of interactions of the protein or enzyme with the lipids, the protein is isolated and purified and then reconstituted with model lipid systems. Hence it is vital that one achieves a reproducible preparation of lipid vesicles. The presence of contaminants will alter the properties of the lipid matrix. Lipids isolated from natural sources such as phosphatidylserine (PS) from bovine brain contain an appreciable fraction of unsaturated lipids. The main source of contamination results from aerobic oxidation of the unsaturated acyl chains. It is also important that all traces of the original solvent in which the lipid stock was dissolved are removed, especially in the case of chlorinated hydrocarbons such as chloroform. The lack of contamination of the lipid system must be confirmed before any results using a lipid vesicle preparation can be accepted.

Fluorescence spectroscopic measurements offer the possibility of determining the presence or absence of such contaminants, since the fluorescence intensity of a suitable chromophore may be particularly sensitive to even small quantities of contamination. In addition, since fluorescence studies can provide important insights into the interactions and molecular dynamics of protein lipid complexes, contamination of the sample may invalidate the results. In our work on the interaction of peptidal hormones with the acidic lipid, PS, we elaborated a fluorescence method of checking the quality of the PS vesicles which we prepared. This method was based on the study of the interaction of a small tripeptide, lysine-tryptophan-lysine (KWK) with PS vesicles reported by Dufourcq and coworkers (1981). This experiment demonstrates the techniques which one should use for the reproducible preparation of PS vesicles and how their quality may be verified by a fluorescence method. A brief review of fluorescence follows in order to provide some basic background to the methods.

Membrane Proteins, ed. by Azzi
©Springer-Verlag Berlin Heidelberg 1986

II. Review of Fluorescence

After a molecule has absorbed a photon of light, it is in an "excited" electronic state. This excited state is usually an excited singlet state of the molecule. This metastable state has a number of processes by which it may deactivate, leading to the original ground state, or to an excited triplet state (intersystem crossing), or to a new molecule as a result of a photochemical process. One of the deactivation processes of the excited singlet state is the emission of a photon of light which is designated fluorescence. Another may be a nonradiative pathway resulting from interactions with solvent or other molecules. A rate constant may be assigned for each of the deactivation processes, which are summarized below.

rate constant
k_r — radiative (fluorescence)
k_p — Product formation
k_{isc} — intersystem crossing to the triplet state
k_{nr} — nonradiative
$k_g(Q)$— quenching.

If a molecule is present which interacts with the excited chromophore in a concentration-dependent manner, this is a special nonradiative deactivation pathway, which is usually designated as a quenching process. The fluorescence efficiency is referred to as the fluorescence quantum yield, $\emptyset_F$, and is defined according to the relationship

$$\emptyset_F = \text{number of fluorescent photons/total photons absorbed.} \qquad (1)$$

It can be shown that the quantum yield of fluorescence, $\emptyset_F$, can be expressed in terms of the rate constants listed above:

$$\emptyset_F = k_r / k_r + k_p + k_{nr} + k_{isc} + k_g (Q). \qquad (2)$$

The quantum yield may be determined by integration of the fluorescence spectrum of the sample and comparing this value with that obtained for a reference compound whose quantum yield is known. In most cases the fluorescence intensity at any wavelength is proportional to the quantum yield, and can be conveniently used in a quantitative manner for analytical purposes. It can be seen from Eq. (2) above, that if one affects any of the rate constants in the denominator, the $\emptyset_F$ will change accordingly. For example, if the interactions between molecules change, the value of k_{nr} may be affected, or if there is a quenching contaminant present, then the $k_g (Q)$ term will be modified.

In some fluorescence experiments the degree of exposure of the chromophore to the solvent, e.g., a tryptophan residue in an enzyme, may be determined by the addition of an external quencher such as acrylamide. The more exposed the residue the more efficient the quenching term, and hence the effect on the fluorescence intensity is greater.

The shape and maximum wavelength of the fluorescence spectrum of a chromophore can be indicative of the polarity of the environment of the chromophore as well as showing special solvent interactions.

The measurement of the fluorescence spectrum and quantum yields are considered to be steady-state information and no dynamic information can be obtained from their measurement alone. If the sample is excited with a short pulse of light, it is possible to measure the decay time of the fluorescence and obtain the lifetime of the excited singlet state. The excited singlet state lifetime, τs, is defined as the reciprocal of the sum of all its deactivation processes.

$$\tau_s = \ 1/k_r + k_{nr} + k_{isc} + k_p + k_g(Q). \tag{3}$$

The measurement of the singlet decay time in combination with the determination of the fluorescence quantum yield allows one to determine the values of some of the individual rate constants providing information on the dynamics of the interactions of the chromophore with its surroundings. It has also been shown that different conformations of a trp in a protein may have more than one fluorescence decay time. This is indicative of a conformational heterogeneity in the protein. Factors affecting the conformational distribution may be studied using fluorescence decay measurements. It is beyond the scope of this brief review of some of the principles of fluorescence to elaborate further on this topic and on the topic of fluorescence anisotropy (which can give information on the dynamics of the motion of protein segments); instead the reader is referred to several excellent reviews (Cundall and Dale, eds, 1983; O'Connor and Philipps 1984; Lakowicz 1983).

Fluorescence spectroscopic methods to study the interactions of proteins and lipids have been used extensively. This is because of the facility of making many of the measurements on standard instrumentation, available at moderate cost. The sensitivity of the technique allows one to measure small concentrations of fluorescent chromophores. Different chromophoric groups or substituents may be selectively probed in a macromolecule, giving the technique a useful specificity. There is a wide range of information available from the measurements, such as: the determination of binding parameters; the environment of the chromophore; its conformational heterogeneity; the exposure to an aqueous environment; the distance between different segments of the molecule, the dynamics of segmental fluctuations of proteins; the interactions with other constituents; and the fluidity of the acyl chain of the bilayer. In the case of proteins, the only fluorescent amino acids are phenylalanine (phe), tyrosine (tyr), and tryptophan (trp), which provide intrinsic fluorescent probes of the protein. Phe has a very low extinction coefficient as well as a low fluorescence efficiency, and hence is virtually neglected in protein studies. On the other hand, tyrosine and tryptophan have higher extinction coefficients near 280 nm and generally have higher fluorescence efficiencies. It is possible to selectively excite trp residues using wavelengths > 290 nm, which makes it possible to monitor its fluorescence properties in proteins. Excitation at 280 nm of proteins containing both tyr and trp permits one to determine interactions between these residues. In the absence of trp, the fluorescence properties of tyr in proteins can be studied. Because of the possibility of selectively monitoring the fluorescence of trp, it has received greater attention than the other aromatic amino acid residues in fluorescence studies.

The fluorescence experiment in this article uses the trp fluorescence spectrum of the tripeptide KWK to confirm that a satisfactory vesicle preparation was obtained. When a trp residue is fully exposed to an aqueous environment as it is for KWK in an aqueous buffer then the spectral maximum is usually found near 350 nm. When this basic peptide (net charge +2) binds to the acidic PS vesicles, the fluorescence spectral maximum undergoes a shift to 335 nm. A significant increase of the fluorescence also occurs with the intensity increasing by a factor of > 1.8 at 335 nm. The interpretation of these changes was made clearer in a study which we performed with KWK and dimyristoylphosphatidylserine (DMPS) vesicles (Yamashita and Szabo, unpublished). A summary of these latter experiments follows and is illustrative of some of the information one may obtain from fluorescence of protein lipid complexes.

DMPS vesicles at pH 5 have a clear gel to liquid crystalline phase transition at 39°C. When DMPS vesicles were added to a solution of KWK at temperatures below the phase transition (25°C), a small shift of the spectral maximum from 350 nm to 348 nm was observed, together with an increased fluorescence of nearly 50% at 348 nm at saturating concentrations of DMPS (Fig. 1) compared to the fluorescence intensity of KWK alone. At temperatures above the phase transition (45°C), as DMPS was added to KWK the fluorescence maximum shifted from 350 nm to 335 nm and there was a doubling of the fluorescence intensity at 335 nm (Fig. 1). When the KWK-DMPS complex, which was prepared at 25°C, was heated to temperatures above the phase transition, with spectra being recorded at different temperatures, the wavelength maximum remained constant at 348 nm until the transition temperature was reached and then the spectral maximum shifted abruptly to 335 nm (Fig. 2). On cooling, the spectral changes were completely reversible. These results indicate that when KWK binds to DMPS vesicles in the gel phase the trp residue is excluded from the lipid matrix remaining exposed to the aqueous environment. In the liquid crystalline phase, when the lipid matrix is more fluid and less densely packed, the indole nucleus of the trp is able to penetrate into the bilayer interior. The increased fluorescence intensity in the gel phase shows

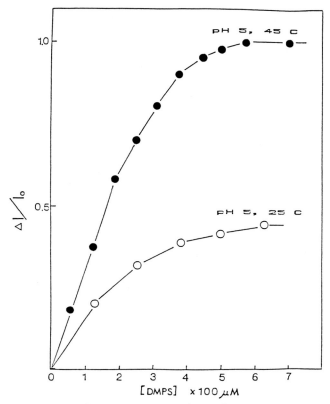

Fig. 1. Relative fluorescence change of KWK fluorescence with added DMPS vesicles

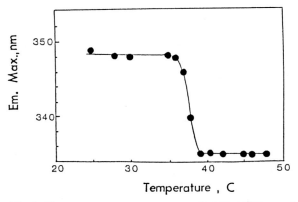

Fig. 2. Fluorescence maximum change of KWK-DMPS complex (1:52) with temperature

that collisional deactivation by the solvent is reduced. An even greater change occurs in the liquid crystalline phase, since the trp residue has been able to enter into the lipid matrix.

The vesicles prepared from bovine brain PS are probably in a disordered liquid crystalline phase owing to the heterogeneity of the acyl chain composition. Hence at the temperature of the experiment (20°C), the trp residue of KWK is able to penetrate into the lipid matrix, resulting in a shift of the spectral maximum to 335 nm and a large increase of the fluorescence intensity. We noted in earlier studies that if the lipid preparation was not carefully lyophilized under nitrogen or argon gas, and all traces of $CHCl_3$ were not removed, or poor quality $CHCl_3$ was used in the dilution of the lipid sample, then no increase in fluorescence intensity was observed although there was a spectral shift indicative of binding of KWK to the PS vesicles. In these cases the contaminants quenched the trp fluorescence.

III. Experimental Procedure

A. Equipment

— Absorption spectrophotometer
— Fluorescence spectrophotometer
— Vortexer
— Ultrasonic bath
— Vacuum pump
— Vacuum desicator
— Rotary evaporator
— Pipets, 1 ml, 2 ml
— Micropipets and pipeter
— Pasteur pipets
— Capillary tubing
— Thin layer chromatography Silica analytical plates 5 cm x 20 cm
— Chromatography tanks
— Round-bottom flasks and stoppers, 50 ml
— Volumetric flask, 5 ml
— Polyfilm
— Aluminium foil
— 1 cm quartz fluorescence cuvettes

B. Chemicals

— Phosphatidylserine (bovine brain) in $CHCl_3$
— Nitrogen gas
— Iodine
— Spectrograde methanol
— Spectrograde chloroform
— Freshly prepared chromatography solutions
      Solution A: $CHCl_3$, CH_3OH, NH_4OH (28%); 65:25:5
      Solution B: $CHCl_3$, CH_3OH, CH_3CO_2H, H_2O; 25:15:4:2
— Buffer, 25 mM sodium cacodylate, pH 5.3, 0.1 mM EDTA
— Lysine-tryptophan-lysine stock solution 1 mg/100 ml in buffer

C. Experiment 1: Analysis and Preparation of Phosphatidyl Serine Vesicles

The purity of the PS must be checked for lysophosphatidylserine (lyso PS) and free fatty acid (FFA) by TLC. The Rf values (Kates 1972) for the two solvent systems are:

Solvent A:	Compound	Rf
	PS	0.05
	lyso PS	0.00
	FFA	0.38
Solvent B:	PS	0.55
	lyso PS	0.15
	FFA	1.00

During the experiment always keep the lipid samples under a nitrogen atmosphere and minimize exposure of the $CHCl_3$ solutions of lipid to light. Open the vial of PS in $CHCl_3$, keeping a nitrogen gas stream flowing over the vial. Transfer this solution to a 5-ml volumetric flask and make up to volume with $CHCl_3$. If there is too much solvent, evaporate the excess using the nitrogen stream. Slowly bubble nitrogen gas through this stock solution for 10 min, keeping the volume constant by the addition of $CHCl_3$. Stopper the flask, and if being stored, wrap the stopper with parafilm and store covered with foil in a freezer. When some of this stock solution is required, allow the solution to come to room temperature prior to opening. When finished, bubble nitrogen through the remaining stock for 2 min and reseal as above.

    The purity of the PS sample is checked by TLC analysis. Using capillary tubing spot each of two silica plates with three spots of PS from the stock solution on a line approximately 1 cm from one edge of the plate. Elute one plate in solution A and the other in solution B until the solvent front reaches a line about 2 cm from the top of the plate. Dry the plates in a fume cupboard and place in the tank containing iodine until the plates darken and spots appear. Remove from the development tank and record the distance between the base line and each spot and the distance between the solvent front and the baseline.

To prepare vesicles of PS, transfer 1 ml of stock PS solution to a clean 50 ml round-bottom flask. Exhaustively evaporate the chloroform for at least 2 h on a rotary evaporator equipped with a vacuum pump. All traces of $CHCl_3$ should be removed, leaving a thin film of dried lipid around the bottom of the flask. Blow nitrogen gas carefully into the flask prior to stoppering it. Before adding the buffer to the film of lipid, bubble nitrogen gas through the buffer solution for 10 min. Add 1.5 ml of buffer to the PS sample, stopper the flask and vortex for about 1 min to suspend all the lipid in the buffer. Place the stoppered flask in a sonicating bath for at least 1 h, keeping the temperature at 30°C during this time. Check for sonication after 1/2 h and 1 h by measuring the absorption spectrum of the sample, which has been diluted by adding 100 μl to 2 ml of buffer, between 400 nm and 250 nm in an absorption spectrophotometer. Use a buffer blank in the reference beam. Sonication is judged to be complete when the diluted solution is clear, the scattered light is very low and the stock solution is translucent. If the sample is still very opaque and has a high scattering efficiency, continue sonication (Sartor and Cavatorta, this Vol.).

After sonication to form vesicles is complete, repeat the TLC analysis to check for any degradation of the PS. Be sure to pass nitrogen gas into the flask with the vesicle preparation while samples are being removed from it. Usually a phosphate determination would be carried after effective sonication has been achieved to measure the PS concentration. However, for these experiments assume that the concentration can be calculated from the concentration of the stock $PS/CHCl_3$ solution. Use a value of 805 g for the molecular weight of the PS.

— Record the results of the TLC analyses
— Retain the light-scattering curves from the absorption measurement.
— Calculate the concentration of PS in the vesicle sample.

D. Experiment 2: Fluorescence Study of the Interaction of Lysine-Tryptophans-Lysine (KWK) with PS Vesicles

The fluorescence experiment uses the trp fluorescence spectrum of KWK to confirm that a satisfactory vesicle preparation was obtained. When all the peptide is bound to the PS, a shift of 15 nm of the fluorescence spectral maximum from 350 nm to 335 nm, and an increase in the fluorescence intensity at 335 nm by a factor of 1.8 should be observed. This is indicative of a satisfactory vesicle preparation.

Pipet 2 ml of KWK stock solution into a fluorescence cuvet and 2 ml of the buffer solution into another cuvet. Gently bubble nitrogen gas through the KWK sample for about 10 min. Stopper both cuvets. The spectrofluorimeter sample compartment should be thermostated at 20°C. The excitation wavelength should be 280 nm. The excitation and emission slits should be set for 4 nm bandpass. Zero the recorder with both the excitation and emission shutter closed. The photomultiplier should have a dynode voltage setting of 800 V working in the ratio mode. Adjust the sensitivity of the instrument amplifier so that a fluorescence reading of 50 is achieved at 350 nm for the sample. Record the fluorescence spectrum from 300 nm to 425 nm of the sample and buffer blank under identical instrument conditions. Keeping the cuvets and PS vesicle preparations under nitrogen, transfer the same aliquots of the sonicated lipid to both cuvets. Measure the fluorescence spectrum after each aliquot addition on the same chart paper as the original spectrum. Aliquots of 25, 50, 100, and 200 μl should be added. Do not adjust the instrument settings during these measurements. Subtract the blank reading from the sample fluorescence intensity at 350 nm and 335 nm.
— Record the wavelength maximum for each spectrum
— Record the ratio of the intensity at 335 nm for the sample with added vesicles to that value obtained for KWK alone. Correct for dilution effects.
— Plot these values against total PS added.

Typical fluorescence spectra of the KWK in absence and presence of a saturating amount of PS vesicles are shown in Fig. 3.

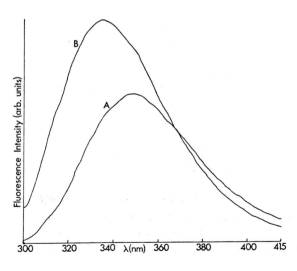

Fig. 3. Fluorescence spectra of KWK; *A* in buffer pH 5.3 alone; *B* with 200 μl PS vesicles added

References

Cundall RB, Dale RE (eds) (1983) Time-resolved fluorescence spectroscopy in biochemistry and biology. Nato Asi Series A, Vol. 69, Plenum, New York

Dufourcq J, Faucon JF, Maget-Dana R, Pileni MP, Helene C (1981) Biochem Biophys Acta 649:67–75

Kates M (1977) In: Laboratory techniques in biochemistry and molecular biology, Vol. 3, pt. II, Work TS, Work E (eds) North Holland, Amsterdam

Lakowicz JR (1983) Principles of Fluorescence Spectroscopy. Plenum, New York

O'Connor DV, Philipps D (1984) Time-correlated single photon counting. Academic Press, London

Purification and Reconstitution of the Proton Translocating Membrane Bound Inorganic Pyrophosphatase from *Rhodospirillum rubrum*

M. BALTSCHEFFSKY and P. NYRÉN

I. Introduction

Photophosphorylation carried out by chromatophores from the photosynthetic bacterium *Rhodospirillum rubrum* will yield both ATP and inorganic pyrophosphate, PPi. The PPi synthesis is coupled to light-induced electron transport in a manner analogous to ATP synthesis (Baltscheffsky et al. 1966) and the enzyme catalyzing this reaction is a H^+-translocating, membrane-bound PPase (Baltscheffsky 1967). The PPi synthesis is a reversible reaction and when PPi is hydrolyzed, the energy released is able to drive a number of energy-requiring reactions. For a recent review on this enzyme see Baltscheffsky and Nyrén (1984). A partly purified preparation of the enzyme (Rao and Keister 1978) has also been shown to act as a PPi-dependent electric generator (Konrashin et al. 1980), when incorporated into a planar phospholipid membrane.

The PPase behaves like an integral membrane protein, spanning the entire chromatophore membrane. Earlier attempts to solubilize and purify this enzyme have only had very limited success (Rao and Keister 1978). Recently we have been able to obtain an apparently pure enzyme with reasonable stability in satisfactory yield (Nyrén et al. 1984), and giving us the possibility to eventually perform a closer characterization of this protein, a characterization which, due to the analogy between the PPase and the ATPase and the apparent comparative simplicity of the former, may also provide closer insight into fundamental aspects of the functioning of coupling factor proteins in general.

II. Method of Solubilization and Purification

A. Preparation of Chromatophores

Rhodospirillum rubrum, strain S-l, is grown anaerobically in the light in batch culture for 40 h at 30°C on the synthetic medium described by Bose et al.

Membrane Proteins, ed. by Azzi
©Springer-Verlag Berlin Heidelberg 1986

(1961). The bacteria are harvested by centrifugation, resuspended in 0.02 M glycyl-glycine buffer pH 7.4 and washed once. The bacterial pellet is resuspended in ice cold 0.2 M glycyl-glycine pH 7.4 in a concentration of about 20 g of bacteria (wet weight) per 100 ml suspension and then disrupted in a Ribi cell fractionator at 20,000 psi, keeping the temperature as low as possible. The cell homogenate is supplied with DNAase and RNAase and then centrifuged for 60 min at 10,000 g in a Servall refrigerated centrifuge. The supernatant from this centrifugation is then recentrifuged for 90 min at 100,000 g in a Spinco preparative ultracentrifuge in order to sediment all membraneous material, the chromatophores. The chromatophores are resuspended and washed twice with 0.2 M glycyl-glycine, pH 7.4 with centrifugation for 60 min at 100,000 g in each wash. The washed chromatophores are finally suspended in a minimal volume of 0.2 M glycyl-glycine and kept in the dark on ice until used. After breaking the cells, all operations are carried out as close to 0°C as possible.

B. Solubilization of the PPase

3 ml of chromatophores (60–70 mg protein ml^{-1}, about 1.5 mM Bchl) are mixed with 21 ml of Tris-HCl buffer, pH 8.4 containing 2.5% (v/v) Triton X-100, 0,75 M MgCl$_2$, 25% ethyleneglycol and 0.2 mM DTE. The suspension is kept on ice and gently stirred for 20 min, after which it is centrifuged for 60 min at 215,000 g. The supernatant containing the solubilized pyrophosphatase is decanted and kept on ice until further use, or alternatively kept frozen at −70°C.

C. Chromatography on Hydroxyl Apatite

6 ml of the solubilized enzyme is desalted on a Sephadex G-25 column, 3.2 x 8.5 cm, equilibrated with 50 mM Tris-HCl, pH 8.4, 25% (v/v) ethyleneglycol, 0.1% Triton X-100, 1 mM MgCl$_2$ and 0.2 mM DTE. This is done in order to reduce the ionic strength of the solution and enable the binding of the enzyme to the hydroxylapatite. 8 g Bio-gel HTP is rehydrated in 50 mM Tris-HCl buffer pH 8.4 containing 0.1% (v/v) Triton X-100, 25% (v/v) ethyleneglycol, 1 mM MgCl$_2$ and 0.2 mM DTE. The rehydrated gel is packed into a 3.2 x 10 cm column and equilibrated with the same buffer. The sample is applied on the column with a flow rate of 1–2 ml per min. After application of the sample, the flow is stopped for 2 min, then the column is washed first with 50 ml of the equilibration buffer with 10 mM MgCl$_2$ then with 30 ml of the same buffer with 130 mM MgCl$_2$. The enzyme is then eluted with 0.2 M MgCl$_2$

in the above described buffer, and fractions of 6 ml are collected. Those fractions showing the highest PPase activity (in a typical experiment fraction 5 to 8) are pooled and concentrated to a final volume of 1 ml by ultrafiltration through an Amicon YM 30 (50 psi) or XM 100 A (25 psi) membrane. This preparation is about 80—90% pure as judged by gel electrophoresis and may be used directly or stored frozen at −70°C. The specific activity of the preparation, before concentration, expressed as micromoles of PPi hydrolyzed in 1 min by 1 mg protein is typically 24, with higher losses with the XM 100 A than with the YM 30 filter. Since the XM 100 A filter allows the passing through of Triton X-100 micelles, the final product has a lower content of detergent and is better suited for incorporation into liposomes than the one concentrated with the YM 30 filter, which on the other hand has higher activity and thus may be better suited for other purposes such as kinetic studies.

D. Incorporation of the PPase in Liposomes

The enzyme preparation, after concentration as described above, is desalted on a Sephadex G-25, (coarse) column, 0.5 x 25 cm. The column is equilibrated with 50 mM Tris-HCl buffer, pH 8.0, 25% (v/v) ethylene glycol, 5 mM $MgCl_2$, 0.2 mM DTE (dithiothreitol) and 0.04% Triton X-100, and the enzyme is eluted with the same buffer.

The freeze-thaw technique of Kasahara and Hinkle (1977) for preparing liposomes has been used with good results. 40 mg soybean phospholipids are supplemented with 1 ml of medium, containing 10 mM Tris-HCl, pH 7.5, 0.5 mM DTE, 0.5 mM EDTA and 0.05% Na-cholate. The suspension is flushed with nitrogen and sonicated in a bath-type sonicator, model G 112 SP 1T, Lab. Supplies (Hicksville, NY). The PPase preparation, 0.2 ml, containing 50—100 μg protein, is then added to 0.3—0.4 ml of liposomes, sonicated for 10 s and rapidly frozen in a dry-ice-ethanol bath. After thawing at room temperature, the preparation is stored on ice.

If liposomes with both the F_oF_1 complex and the PPase are desired, the procedure is the same as above, with the exception that together with the PPase preparation, 0.1 ml containing 10—20 μg protein, is added 0.1 ml of a purified ATPase preparation (50 μg protein), prepared according to Pick and Racker (1979) with some minor modification (Nyrén and Baltscheffsky 1982). Figure 1 shows ATP synthesis driven by PPi hydrolysis in such liposomes.

In the liposomes the PPase has regained certain characteristics, typical for proton pumping activity, such as stimulation of hydrolysis by uncouplers or ionophores (Shakov et al. 1982).

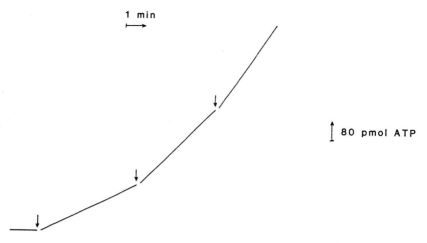

1 min

80 pmol ATP

Fig. 1. Time course of ATP synthesis in a suspension of PPase, ATPase liposomes; 25 μl liposomes (0.07 mg protein ml^{-1}) were suspended in a medium containing 1 ml 0.2 M glycylglycine (pH 7.8), 0.2 ml luciferin/luciferase assay, 20 μl 100 mM sodium phosphate, 10 μl 10 mM ADP and 50 μl 10 mM sodium pyrophosphate; final [MgCl$_2$] was 10 mM; ($\rightarrow$) 25 μl liposomes added. The resulting luminescence was measured in an LKB luminometer 1250. The light output was calibrated by addition of a known amount of ATP. (Nyrén and Baltscheffsky 1983)

E. Determination of PPase Activity

The PPase activity is assayed in a reaction mixture containing 0.75 mM MgCl$_2$, 0.5 mM Na$_4$P$_2$O$_7$, 1 ml 0.1 M Tris-HCl, pH 7.5, 0.2 mg ml^{-1} asolectin or 25 μg ml^{-1} cardiolipin, when required, and H$_2$O, in a total volume of 2 ml. Asolectin is prepared by sonication in 0.1 M Tris-HCl, pH 7.5, until a clear suspension is obtained. The assay mixture is incubated at 30°C and the reaction is terminated after the desired time by addition of 1 ml 10% trichloroacetic acid. In the blanks the trichloroacetic acid is added before the enzyme sample. If, as sometimes is the case with crude preparations, a protein precipitate forms, the mixture is chilled and centrifuged, and the supernatant fluid is collected. 0.4 ml of the assay mixture or the supernatant is analyzed for Pi according to Rahtbun et al. (1969). This method is reliable, provided that the remaining concentration of PPi is lower than 0.35 mM. At higher concentrations the sensitivity is gradually lost, and misleading results may be obtained if special attention is not given to the remaining PPi concentration.

F. Polyacrylamide Gel Electrophoresis

Native gel electrophoresis can be performed with desalted enzyme on 5% gels prepared according to Davis (1964), with the exception that 20% ethyleneglycol (v/v), 0.1 mM DTE and 0.1% Triton X-100 were included in the gel. The electrolyte buffer contains 0.1% Triton X-100. The pyrphosphatase activity may be localized by incubating the gel in a solution identical to that used for assaying the enzyme activity for 20 min at 30°C, after which it is rinsed with distilled water and immediately immersed in the triethylaminemolybdate reagent described by Sugino and Miyoshi (1964), which specifically precipitates Pi. A sharp, discrete zone, corresponding to the enzyme activity in the gel, appears within a few min.

In order to elucidate the subunit composition of the enzyme, SDS polyacrylamide gel electrophoresis, essentially according to Weber and Osborn (1969), has been used. To dissociate the enzyme into subunits, 1 part of purified enzyme solution is incubated overnight at 37°C with 1 part of 0.25 M Tris-HCl, pH 6.9, 20% glycerol, 2% SDS and 2% merkaptoethanol. The bands are stained with Coomassie brilliant blue. This procedure yields 6–7 discrete bands with apparent molecular weights of respectively 64, 52, 41, 31 (25), 20, and 15 kDa. The 25 kDa band is usually very weak or absent and may not be part of the enzyme. Also, it cannot at present be excluded that some of these bands are dimers of others, especially since the protein is rather difficult to dissociate.

G. Substrate Specificity

The hydrolyzing activity of the PPase is very specific for PPi. The activity with a number of other compounds containing the pyrophosphate moiety, ATP, ADP, IDP, MDP, tetrapolyphosphate, and 3-glycerophosphate, is zero. With tripolyphosphate there is a low activity. In the case of tri- and tetrapolyphosphate there is an apparent initial activity which seems to be due to PPi contamination in the reagents. Time curves of the PPase activity with these substrates show that the activity ceases after 10–30 min, whereas the activity with PPi is more or less linear for 80 min. With tripolyphosphate the low remaining activity is 6% of that with PPi. Imidodiphosphate and methylenediphosphonate act as competitive inhibitors of PPi hydrolysis.

H. Inhibitors of the PPase Activity

Table 1 shows the action of a number of compounds on the PPase activity, both when the enzyme is membrane bound, either to the chromatophores or

Table 1. Effect of some substances on membrane-bound an purified inorganic pyrophosphatase. Particles corresponding to 120 μg protein and purified enzyme corresponding to 15 μg protein was assayed as described in the text except for the MDP and IDP treatment for which 12.5 mM $MgCl_2$ was used instead of 0.75 mM. The DCCD, MalNET Nbf-Cl treatment was performed by incubation of particles and enzyme for 10 min at 0°C or 30°C. Dio-9 was incubated with particles and enzyme for 20 min at 0°C. The inorganic pyrophosphatase reaction was initiated by adding PPi (Nyrén et al. 1984)

Additions	Concentration		Activity of inorganic pyrophosphatase dependent on	
			Membrane-bound	Purified
			% of control	
FCCP	1.5	μM	150	100
DCCP (0°C)	100	μM	30	97
NaF	5	mM	83	32
	10	mM	64	11
	20	mM	45	4
MDP	0.1	mM	84	64
	0.2	mM	74	36
IDP	0.1	mM	48	24
	0.2	mM	34	12
NBF-Cl (0°C)	0.25	mM	33	23
	0.50	mM	18	7
NBF-Cl (30°C)	0.25	mM	82	12
	0.50	mM	61	1
N-methylmaleimide (0°C)	1	mM	30	20
	2	mM	13	13
N-methylmaleimide (30°C)	1	mM	99	84
	2	mM	85	76
Dio-9 (0°C)	15	μg ml^{-1}	80	40
	30	μg ml^{-1}	60	30

incorporated into liposomes, and in the solubilized state. Classical uncouplers and ionophores, as has been mentioned before, are only effective with the membrane-associated enzyme, whereas inhibitors which can be assumed to act at, or close to, the catalytic site are effective both with the membrane-bound and the solubilized enzyme. It is also interesting to note that the inhibitors which may be assumed to act at or near the catalytic site are usually much more efficient with the solubilized enzyme, which may indicate that this site in the original chromatophore membrane resides somewhat buried below the membrane surface.

References

Baltscheffsky H, von Stedingk LV, Heldt HW and Klingenberg M (1966) Science 153: 1120–1122

Baltscheffsky M (1967) Biochem Biophys Res Commun 28:270–276

Baltscheffsky M and Nyren P (1984) In: Ernster L (ed) Bioenergetics. Elsevier Science Publ. Amsterdam, pp 187–197

Bose SK, Gest H and Ormerod JG (1961) J Biol Chem 236:13–19

Davis BJ (1964) Ann NY Acad Sci 121:404–412

Kasahara M and Hinkle P (1977) J Biol Chem 252:7384–7390

Kondrashin AA, Remennikov VG, Samuilov VD and Skulachev VP (1980) Eur J Biochem 113:219–222

Nyren P and Baltscheffsky M (1983) FEBS Lett 155:125–130

Nyren P, Hajnal K and Baltscheffsky M (1984) Biochim Biophys Acta 766:630–635

Pick U and Racker E (1979) J Biol Chem 252:2793–2799

Rao PV and Keister DL (1978) Biochem Biophys Res Commun 84:465:473

Rathbun WB and Betlach WM (1969) Analyt Biochem 28:436–445

Shakov YA, Nyrén P and Baltscheffsky M (1982) FEBS Lett 146:177–180

Sugino Y and Miyoshi Y (1964) J Biol Chem 239:2360–2369

Weber K and Osborn M (1969) J Biol Chem 244:4406–4412

The Function of Transmembrane Channels: Ion Transport Studies by ^{23}Na NMR

A. SPISNI, V. COMASCHI and L. FRANZONI

I. Introduction

A. Relevance of Ion Movement

It is well recognized that energy transfer in biological processes is closely coupled with ionic gradients across the lipid bilayers of cell membranes. Furthermore, the selectivity of ionic permeation of biomembranes is fundamental to cellular excitability.

There are two major mechanisms by which ion transport and selectivity can be achieved, namely carriers and transmembrane channels. Carriers are macromolecules able to selectively coordinate ions and then dissolve in the lipid moiety moving from one side to the other of the membrane where, following decomplexation, they release the ions. On the other hand, transmembrane channels are macromolecules able to form pores that span the lipid bilayers offering to the ions an hydrophilic pathway through which they can be translocated.

B. The Gramicidin A (GA) Channel

Gramicidin A is a polypentadecapeptide isolated from *Bacillus brevis,* HCO-L-Val$_1$-Gly$_2$-L-Ala$_3$-D-Leu$_4$-L-Ala$_5$-D-Val$_6$-L-Val$_7$-D-Val$_8$-L-Trp$_9$-D-Leu$_{10}$-L-Trp$_{11}$-D-Leu$_{12}$-L-Trp$_{13}$-D-Leu$_{14}$-L-Trp$_{15}$-NCH$_2$CH$_2$OH, and it was the first transmembrane channel to be structurally described (Urry 1971, Urry et al. 1971).

It has been demonstrated that when incorporated in a lipid phase it assumes a helical conformation that has been proposed to be a β-left-handed helix. Moreover, data have been reported suggesting that the channels are formed by N-terminal dimers (Masotti et al. 1980, Weinstein et al. 1980, Wallace et al. 1981). Finally, its selectivity toward Na$^+$ ions is achieved by the ability of the peptide's carbonyls to effectively coordinate the ions throughout the channel (Urry 1973, Sandbloom et al. 1978).

Membrane Proteins, ed. by Azzi
© Springer-Verlag Berlin Heidelberg 1986

II. ^{23}Na NMR

As we are interested in the study of the interaction of Na$^+$ ions with macro-molecules, the choice of a technique such as ^{23}Na NMR seems to be the most obvious. The direct analysis of the ^{23}Na magnetic resonance has great advantages over the other methods normally used to study the behavior of Na in biological systems. First of all, the ^{23}Na NMR spectra are very simple as they are generally characterized by a single resonance (only in special cases can two lines be observed). Second, the width of the steady-state ^{23}Na nuclear magnetic resonance is very sensitive to the modifications of the chemical and physical environment of the Na$^+$ ions. As a consequence, from the analysis of the linewidth we can expect to obtain valuable information on their coordination state.

A. The Relaxation Mechanism for Quadrupolar Nuclei

A brief account of the origin of an NMR signal has been reported in part 1 of the theory section of the chapter *Polypeptide-lipid interactions as studied by ^{13}C NMR* (Spisni et al. Part II, this Vol.). Attention has to be drawn to the fact that this nucleus possesses an electric quadrupole moment and to the consequences of that on the nuclear relaxation. An electric quadrupole moment arises from a nonspherical distribution of the electrical charges at the nucleus and can be viewed as the combination of two electric dipoles displaced from one another and pointing in opposite direction. This feature is characteristic of all nuclei with a nuclear spin $I > 1/2$: ^{23}Na has I=2/3.

For the electric quadrupole moment to become effective in the relaxation mechanism of the nucleus, it is necessary to interact with an electric field gradient. In fact, the quadrupolar relaxation is the only one that is triggered by electrical interactions rather than by magnetic ones.

In the extreme narrowing limit that, for standard magnetic fields, as we have shown in the experiment on *Polypeptide-lipid interactions as studied by ^{13}C NMR* (Spisni et al. Part II, this Vol.) corresponds to molecular motions faster than 10^{-8}s, the contribution of the electric quadrupole moment to the nuclear relaxation can be written as:

$$\frac{1}{T_1} \simeq \frac{1}{T_2} = \frac{3}{40} \cdot \frac{2I+3}{I^2(2I-1)} \left(1 + \frac{\eta^2}{3}\right)\left(\frac{e^2\,qQ}{\hbar}\right)^2 \tau_c, \qquad (1)$$

where T_1 and T_2 are the longitudinal and transversal relaxation times, I is the nuclear spin, η is the asymmetry parameter, $\left(\frac{e^2qQ}{\hbar}\right)$ is the quadrupolar coupling constant (QCC) and τ_c is the correlation time.

Generally, the quadrupolar mechanism dominates over the other relaxation mechanisms, leading to broad resonance lines. As the QCC can vary from very small values to hundreds of MHz, if we consider a nucleus with spin I= 3/2, $\eta = 0$ and with a QCC of 5MHz, from Eq. (1) we obtain $T_1 = 10^{-14}$ $(\tau_c)^{-1}$. For a molecule tumbling in a liquid, as $\tau_c = 10^{-12}$s, we end up with a value of T_1 in the range of 10^{-2}s, that is definitely very short, and that will give rise to broad lines. However, apart from the influence of τ_c, the linewidth is also affected by other factors. Some of these factors are strictly dependent on the intrinsic characteristics of the nucleus itself, such as the quantum number I and the quadrupolar moment Q, so that nuclei with high spins show sharper lines than the ones with low spin. Others, instead, depend on the nature of the surroundings such as the local symmetry, at the nucleus, of the electric field gradient. In fact, any site of tetrahedral, octahedral, or spherical symmetry will have, in principle, zero field gradient and the expected resonance line will be sharp as the quadrupolar relaxation is not effective. In the case of simple ions in aqueous solution such as Na^+ or of the ^{14}N in $[NMe_4]^+$, in deed it is possible to detect reasonably sharp lines. However, even in such cases, long-range perturbations or temporary interactions such as the formation of ion pairs or of other complexes may perturb the symmetry at the nucleus and activate the quadrupolar relaxation mechanism, thereby producing a broadening of the resonance line. In conclusion, the measure of the linewidth for these nuclei may provide valuable information in the study of those processes.

B. The Measure of the Spin-Lattice Relaxation

As $\Delta\nu_{1/2}$, the linewidth at half height (see Fig. 1) is related to T_2 by the equation:

$$\Delta\nu_{1/2} = \frac{1}{\pi T_2} \tag{2}$$

In the extreme narrowing condition, where $T_1 = T_2$, an estimate of T_1 can be obtained from the linewidth analysis. Assuming that during the process under investigation the QCC does not change appreciably, any broadening of the linewidth can be ascribed to a parallel increase of the correlation time (slowing down of the molecular motion) and/or to a perturbation of the electric field gradient symmetry at the nucleus (formation of ion pairs or of other complexes) that enhance the efficiency of the quadrupolar relaxation.

Another method that allows a direct determination of the spin-lattice relaxation rate is to measure the time requested for the macroscopic magnetization M_0 to return to equilibrium after it has been totally inverted. This method requires a two-pulse sequence as shown below: (Fig. 2).

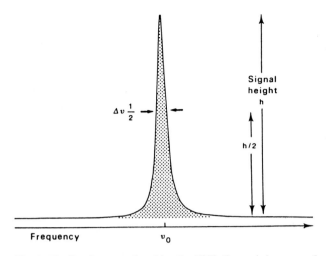

Fig. 1. The lineshape predicted by the NMR theory is known as Lorentzian. In the extreme narrowing limit, where $T_1 = T_2$, the width of the absorption line at half height is equal to $(\pi T_1)^{-1}$

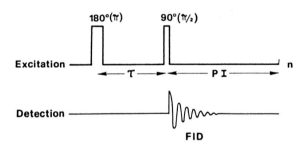

Fig. 2. Diagram of the multi-pulse sequence used in the determination of the spin-lattice relaxation time, T_1, with the inversion recovery method

The 180° (π) pulse produces an inversion of the magnetization M_0 to $-M_0$ and a variable time is allowed during which the magnetization begins to relax back to equilibrium via spin-lattice relaxation. A 90° ($\pi/2$) pulse is then applied to "read" the state of the magnetization at that specific time τ. It can be demonstrated that the magnetization obeys the following equation:

$$M(t) = M_0 \left[1 - 2 \exp\left(-\tau/T_1\right)\right]. \tag{3}$$

Equation (3) can be rewritten as:

$$M(t) - M_0 = 2M_0 \exp\left(-\tau/T_1\right), \tag{4}$$

therefore if we plot the natural logarithm of the peak height against τ, T_1 can be derived as the slope of the line described by the equation:

$$\ln [M_0-M(t)] = \ln 2 + \ln M_0 - \tau/T_1. \tag{5}$$

Moreover, there will be a certain interval of time τ_{null} at which the magnetization $M(t) = 0$, for that value of τ_{null}, Eq. (5) will become: $T_1 = \tau_{null}/\ln 2$. This value of τ_{null} therefore, can be used to derive a rough estimate of the spin-lattice relaxation time.

In the specific case of the GA incorporated in the lysolecithin (LY) phase, it has been shown that the analysis of the behavior of T_1 and T_2 as a function of the Na^+ concentration has allowed the determination of the binding constants and of the kinetic constants necessary to describe the movement of the Na^+ ions through the channel (Urry et al. 1980a,b).

III. Aim of the Experiment

The aim of this experiment will be to show that, following the behavior of the linewidth and of the T_1 of the ^{23}Na magnetic resonance during the various steps of the incorporation of GA into the LY phase, it is possible to demonstrate that indeed the GA forms channels which interact specifically with the Na^+ ions.

In fact, in the presence of channels through which the Na^+ ions can move via coordination at the peptide carbonyls (Urry 1973), then a reduction of the mobility and possibly a perturbation of the symmetry of the electric field gradients at the Na^+ ions is expected. This event, therefore, ought to induce an increase of the relaxation rate and a consequent broadening of the linewidth.

The specificity of the interaction of the Na^+ ions with the GA channels will be verified by adding to the suspension Ag^+ ions that are known to block the Na^+ transport. If the observed broadening is specifically due to the interaction with the active channels, the experimental data should show that after the addition of a blocking agent such as Ag^+, the Na^+ ions are prevented from interacting with the binding sites inside the channels, and as a consequence both the linewidth and the T_1 should change back to the values measured for the Na^+ free in aqueous solution.

IV. Experimental Procedures

A. Reagents

— Lysolecithin (LY)
— Gramicidin A (GA)
— NaCl 100 mM
— AgNO$_3$ 100 mM
— D$_2$0 98%

B. Equipment

— NMR Spectrometer
— Sonifier equipped with microtip
— Thermostatable bath
— Vortex mixer
— VSL pyrex tubes with caps
— 10 mm NMR precision tubes

C. Preparation of the Samples

For the preparation of the samples follow the description of paragraph 3) of
the experiment *Lipid-protein interactions as studied by* ^{13}C *NMR* (Spisni
et al. Part II, this Vol.).

D. NMR Experiments

a) Transfer each sample in a 10-mm NMR precision tube and set up the
Spectrometer at the frequency for ^{23}Na, that at a field of 4.7 T is about 52.90
MHz. Working with the quadrature detection, once the resonance line of the
Na$^+$ in water has been found, set the offset frequency in order to have the signal in the center of the oscilloscope. Choose the sweep width so that the aqui-
sition time is longer than about 200 ms. As the relaxation of the Na$^+$ ions in
water is normally not longer than 50 ms, a pulse interval as short as 100–150
ms can be used.
b) Determine the linewidth and the T$_1$ for the ions in water and in the presence of LY miscelles (sample a). The results should be similar. Use these values
as a reference.
c) Determine the linewidth and the T$_1$ for the other two samples (b,c).

d) After the spectrum of the sample containing the GA channels has been collected, add an amount of Ag^+ equivalent to the Na^+ present and repeat the measurement. Now the linewidth and the T_1 values should be similar to the reference ones.

References

Masotti L, Spisni A, Urry DW (1980) Cell Biophys 2:241—251

Sandbloom J, Naher E, Eisenmann G (1978) J Membr Biol 40:97

Urry DW (1971) Proc Natl Acad Sci USA 68:672—676

Urry DW (1973) In: Conformation of biological molecules and polymers. Jerusalem Symp Quant Chem Biochem 5:723

Urry DW, Goodall MC, Glickson JD, Meyers DF (1971) Proc Natl Acad Sci USA 68: 1907—1911

Urry DW, Venkatachalam CM, Spisni A, Laüger P, Khaled MA (1980a) Proc Natl Acad Sci USA 77:2026—2032

Urry DW, Venkatachalam CM, Spisni A, Bradley RJ, Trapane TL, Prasad KU (1980b) J Membr Biol 55:29—51

Wallace BA, Veatch WR, Blout ER (1981) Biochemistry 20:5754—5760

Weinstein S, Wallace BA, Morrow JS, Veatch WR (1980) J Mol Biol 143:1—19

Membrane Signal Transduction via Protein Kinase C

C.W. MAHONEY, R. LÜTHY, and A. AZZI

I. Introduction*

Protein kinase C (PKC) is thought to play an important role in the signal transduction across the cell membrane in different cell types (Nishizuka 1984). Many different activators act via this pathway. PKC is activated by diacylglycerols that are produced by a phospholipase C as a consequence of the binding of an activator to a receptor. The activated PKC then phosphorylates intracellular substrates, which lead to the responses of the cell to the extracellular activator (Fig. 1). On the other hand, purified PKC binds phorbol esters, substances which have been known for many years as tumor promoters and non-physiological cell activators (Hecker and Schmidt 1974). PKC is a Ca^{2+}-activated, phospholipid-dependent enzyme and is modulated by phorbol esters and diacylglycerols: they activate PKC by lowering the amount of Ca^{2+} and phosphatidylserine needed for maximal activity (Castagna et al. 1982). The binding of radioactive phorbol esters can be inhibited competitively by diacylglycerols (Sharkey et al. 1984, Sharkey and Blumberg 1985), suggesting that diacylglycerol and phorbol esters have the same binding site.

Protein kinase C, although detectable in virtually all tissues examined to date, is most abundant in brain, spleen, platelets, and lymphocytes (Kikkawa et al. 1983, Schatzman et al. 1983, Kuo et al. 1980) and may be purified to homogeneity from rat brain (Kikkawa et al. 1983) or from pig spleen (Schatzman et al. 1983). However, in both cases the procedures are long (2 weeks preparation) and yield only hundreds of micrograms of pure PKC kg^{-1} tissue. In addition, in the case of rat brain, the enzyme is stable for only $1-2$ days after purification. Partially purified PKC may be obtained from bovine brain

*Abbreviations: PS, phosphatidylserine; EGTA, ethyleneglycolbis-(2-aminoethyl)-tetra-acetic acid; EDTA, ethylenediaminotetraacetic acid; ATP, adenosine-5'-triphosphate; TCA, trichloroacetic acid; NaPPi, sodium pyrophosphate; BSA, bovine serum albumin; PKC, protein kinase C; DAG, diacylglycerol; PDB, phorbol-12, 13-dibutyrate; PEG, polyethyleneglycol; TLC, thin layer chromatography; CBB-G, Coomassie brilliant blue G; SOD, superoxide dismutase; DMSO, dimethylsulfoxide; DW, bi-distilled water; kD, kilodalton.

Membrane Proteins, ed. by Azzi
© Springer-Verlag Berlin Heidelberg 1986

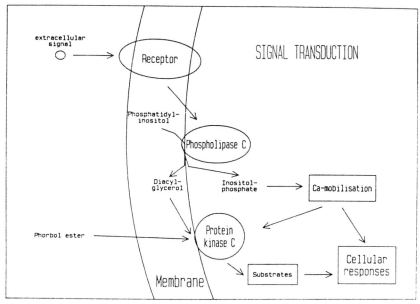

Fig. 1. Signal transduction (see text)

using Phenyl Sepharose and DEAE Sephacel chromatography (Walsh et al. 1984). This preparation contains no other protein kinases or phosphatases.

Crude extracts of homogenized tissue can be used as a PKC source, however it is difficult to show Ca^{2+} and phospholipid dependence in the presence of other protein kinases, residual phospholipid, and Ca^{2+}. Because of the redistribution of PKC from the membrane bound to the cytosolic fraction in the absence of free Ca^{2+}, it is advantageous to homogenize tissue in high levels of divalent metal chelator (i.e., 2 mM EDTA) and to isolate cytosolic PKC. High divalent metal chelator concentrations also obviate the prevalent proteolytic breakdown of PKC by a Ca^{2+}-dependent protease. In the presence of micromolar-millimolar free Ca^{2+}, PKC is readily cleaved into a 50 kDa Ca^{2+} – and phospholipid – independent catalytic and a 30-kDa fragment (Kikkawa et al. 1983, Inoue et al. 1977).

Neutrophil granulocytes are a good model for cell activation. They are activated physiologically by chemotactic peptides or opsonized particles, having a large repertoire of responses (phagocytosis, superoxide production, enzyme release). In vitro they can also be activated with phorbol esters or diacylglycrols to produce superoxide.

II. Materials and Methods

A. Isolation of Neutrophils (Babior and Cohen 1981)

Equipment
— Preparative refrigerated centrifuge

Solutions needed
— ACD: Na citrate   75 mM
            Citric acid   38 mM
            Glucose   124 mM
— 0.9% NaCl
— 0.6 M KCl
— Dextran solution: 6% (w/v) Dextran MW 70,000 (Pharmacia)
                       0.9% NaCl
— Dulbecco's phosphate-buffered saline (PBS):
  NaCl   137 mM
  KCl   2.7 mM
  Na_2HPO_4 8.1 mM
  KH_2PO_4 1.5 mM
  $CaCl_2$ 0.9 mM
  $MgCl_2$ 0.5 mM

Neutrophils are activated by exposure to glass, hence they should be kept in siliconized glassware or plasticware. The starting material can either be anti-coagulated blood or a buffy coat. ACD is used most often as an anticoagulant at a ratio of blood to ACD of 5:1.

2 vol of ACD-blood are mixed with 1 vol of the dextran solution in a plastic cylinder and allowed to stand undisturbed for 1 h at room temperature. The dextran agglutinates most of the red blood cells, which sink, leaving an upper layer of leucocyte-rich plasma.

This upper layer is carefully removed and placed on ice. Then the cells are pelleted by centrifugation at 100 g for 12 min at 4°C. The supernatant is discarded.

The pellet still contains residual red cells. These are lyzed by the following treatment (hypotonic shock): The pellet is resuspended gently in ice-cold water (1/10 of the original volume of the leucocyte-rich plasma) with a Vortex mixer. Exactly 30 s after the addition of water, the tonicity is restored with 3 vol ice-cold 0.6 M KCl.

The mixture is centrifuged at 160 g , 4 min at 4°C and the pelleted cells are washed by resuspension of the pellet in cold PBS and subsequent centrifugation (160 g, 4 min, 4°C). If the pellet is still red, the steps of lysis and washing are repeated.

The white pellet is resuspended in a small volume of PBS and stored at 0°C.

The cells are counted in a Neubaurer counting chamber under the micros-cope and the suspension diluted to approximate 5×10^7 cells ml^{-1} in PBS and stored at 0°C.

A volume of 100 ml blood yields about 10^8 cells, 90% of those are neutro-phils.

B. Activation of Neutrophils (Babior and Cohen 1981)

Equipment
— Thermostated recording spectrophotometer

Solutions needed
— PBS	as above, 37°C
— Suspension of neutrophils	5×10^7 cells ml^{-1}
— Superoxide dismutase (SOD)	1 mg ml^{-1} in water (Sigma)
— Horse heart cytochrome c	12.5 mg ml^{-1} in PBS (Sigma)
— Phorbol-12, 13-dibutyrate (PDB)	0.4 mM in ethanol or DMSO (Sigma)

CAUTION   Phorboldibutyrate is a tumor-promoter. Avoid skin contact.

1. Principle

Activated neutrophils produce superoxide radicals (O_2^-). The O_2^- produc-tion can be followed by measuring the superoxide dismutase (SOD) inhibi-table cytochrome c reduction. Since SOD destroys specifically O_2^-, the re-duction of cytochrome c by O_2^- can be distinguished from the reduction by other agents. The reduction of cytochrome c is monitored by measuring the absorbance at 550 nm or the absorbance difference $abs_{550nm} - abs_{540nm}$.

2. Procedure

100 μl neutrophil suspension are mixed with 1.7-ml PBS in a cuvette and placed into the holder of the spectrophotometer which is thermostated at 37°C. After 2 min 100 μl of the cytochrome c solution are added and recor-ding is startet. The neutrophils are then activated by adding 10 μl PDB and after 2 min 10 μl of SOD are added to stop the reduction of cytochrome c by O_2^- (Fig. 2).

The amount of reduced cytochrome c can be calculated using the following extinction coefficients:

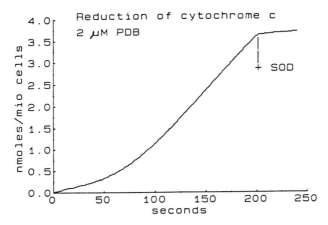

Fig. 2. Cytochrome c reduction by neutrophils activated with 2 μM PDB

$$\epsilon_{550-540} = \quad 19 \text{ mM}^{-1} \text{ cm}^{-1}$$
$$\epsilon_{550} \quad = \quad 29.5 \text{ mM}^{-1} \text{ cm}^{-1}$$

3. Suggestions

A diacylglycerol, for example diolein (Sigma) or oleoyl-acetyl-glycerol (Cal-biochem) can be used to activate the cells instead of phorboldibutyrate. The concentration of activator can be varied and a plot of the curve) vs. activator concentration can be made to estimate the saturating concentration.

C. Preparation of Crude Extract Containing PKC (Kuo et al. 1980)

Equipment
— A refrigerated preparative centrifuge.
— Extraction buffer (4°C): 20 mM Tris-HCl, pH 7.5
                                          2 mM EDTA
                                          50 mM 2-mercaptoethanol
All steps are performed at 4°C.

Rat spleen or brain is homogenized in 3 vol of extraction buffer. The homogenate is centrifuged (30,000 g, 20 min) to remove cell debris and the supernatant is diluted with extraction buffer to give a protein concentration of 1.25–12.5 mg ml^{-1}.

D. Protein Assay (Bradford 1976, Spector 1978)

Stock solutions
— Concentrated dye solution:
    Dissolve 200 mg Coomassie brilliant blue G (CBB-G) in 100 ml 95% ethanol. While stirring add 200 ml conc. H_3PO_4 and then 100 ml DW Store at $4^{o}C$.
— Working dye solution (good for 2 weeks):
    Dilute concentrated dye solution 1:4 with DW and refrigerate for 2 h. Filter the solution through Whatman 1 paper and store at $4^{o}C$.

Assay
Add the working dye solution to the protein sample, vortex, and read the absorbance at 595 nm within 2—15 min. Use the following volumes for the specified protein range needed:

Protein range	Sample volume	Dye volume
0.5—5 μg	0.1 ml	0.7 ml
1—10 μg	0.1 ml	1.0 ml
10—100 μg	0.1 ml	5.0 ml

Make a standard curve using a 1 mg ml^{-1} BSA solution [ϵ_{280} = 0.66 (ml mg^{-1}) cm^{-1}]. Note: SDS interferes with this assay and the Biorad brand of CBB-G gives the best color yield among the dyes tested from several manufacturers.

E. Purity Analysis of Phosphatidylserine by TLC

Merck silica 60 or silica 60F (5 x 20 cm) on glass TLC plates are used. Phosphatidylserine (1—100 μg loads) in $CHCl_3$ are spotted on the TLC plate 2.5 cm up from the bottom edge. The plate is air-dried and then developed in $CHCl_3$/MeOH/28%NH_3 (69:26:5, by vol). The plate is air-dried in a hood and the phospholipids are detected by placing the plate in a chamber with iodine crystals. Within several minutes yellow spots are detectable. PS and diolein have R_f's of 0.10 and 0.94 respectively under these conditions. The use of an alternative solvent system, benzene/diethyl ether/ethyl acetate/acetic acid (80:10:10:0.2), results in R_f's of 0 and 0.58 for PS and diolein respectively. High purity PS is obtained from Avanti Polar Lipids (Birmingham, AL, USA) or lipid Products (South Nutfield, Surrey, UK).

F. Phosphorylation Assay

Equipment
— Water bath at 30°C
— Vacuum manifold for 25 mm filter discs
— Millipore HA (0.45 μm) or Whatman GFB (1.0 μm) filter discs (25 mm diameter)

Stock solutions
— 0.5 M Tris-HCl, pH 7.5 (30°C)
— 25 mM Mg acetate
— 0.25 mM (γ $-^{32}$P) ATP (100,000 cpm nmol^{-1})
— 5 mg ml^{-1} histone (lysine rich fraction, Sigma Type III-S)
— 200 mM $CaCl_2$
— 10 mg ml^{-1} phosphatidylserine (PS) in $CHCl_3$
— 4.9 mM EGTA in 20 mM Tris-HCl, pH 7.5 (30°C)
— 12% trichloroacetic acid (TCA), 2% sodium pyrophosphate (NaPPi) (4°C)
— 6% trichloroacetic acid, 1% sodium pyrophosphate (4°C)
— 10 mg ml^{-1} bovine serum albumin (BSA)
— PKC enzyme
— 0.5 mM ATP (ϵ_{259} = 15,400 M^{-1}cm^{-1})
— 20 mM Tris-HCl, pH 7.5 (30°C)

Final concentration of assay mix (0.25 ml total volume)
— 20 mM Tris-HCl, pH 7.5 (30°C)
— 5 mM Mg acetate
— 10 μM (γ $-^{32}$P) ATP (100,000 cpm nmol^{-1})
— 0.2 mg ml^{-1} histone
— +/− 0.5 mM $CaCl_2$
— +/− 0.04 mg ml^{-1} PS
— +/− 1 mM EGTA (optional)
— +/− 0.1−1.0 μg ml^{-1} diolein (optional)

The amount of PS needed for an experiment is dispensed into a glass tube and dried down with a stream of N_2. To this is added 50 μl 20 mM Tris-HCl, pH 7.5 per assay, and the mixture is sonicated at high power under N_2 and on ice (30 s, twice). The sonicated mixture is then added to the balance of the assay mix. A stock solution of (γ $-^{32}$P)ATP (0.25 mM; 100,000 cpm nmol^{-1}) is made by combining 0.5 vol 0.5 mM ATP (cold) with 0.489 vol DW and 0.0114 vol 1 mCi ml^{-1} (γ $-^{32}$P) ATP (Amersham PB.132 in ethanol/ 1:1) and can be stored frozen in small aliquots. Radioactive work must be done behind Plexiglas shielding and with the use of gloves, lab coat, and safety glasses to protect against the hard radiation of ^{32}P.

To start the reaction 20 μl of PKC enzyme (1.25–12.5 mg ml^{-1} protein) is added to 230 μl of assay mix and the tube is gently mixed and incubated at 30°C for 15 min. The reaction is stopped by the addition of 0.35 ml cold 12% TCA, 2% NaPPi and 100 μl 10 mg ml^{-1} BSA. After mixing, the contents are transferred quantitatively to an HA or GFB filter on a vacuum manifold. The empty tube is washed with cold 6% TCA, 1% NaPPi, and the contents transferred to the filter. The filter is then extensively washed (4–5 portions, several ml each) with cold 6% TCA, 1% NaPPi, and the filter is transferred to a minivial to which 5 ml of scintillation cocktail is added. After wiping the vials with a wet Kimwipe (eliminates potential electrostatic artifactual counts), the minivials are placed into carrier vials and counted in a liquid scintillation counter.

Ca^{2+}-PS-stimulated activity is determined by subtracting the activity in the presence of Ca^{2+} (absence of PS) from that in the presence of both Ca^{2+} and PS. Ca^{2+} stimulated activity (in the absence of PS) accounts for 20–25% and 10–15% for brain and spleen, respectively, of total activity (in the presence of PS and Ca^{2+}).

Phosphorylation specific activities can be determined by performing the above assay on a time course basis (0–10 min) and by measuring the protein content of the PKC source.

To ascertain the presence of Ca^{2+}– and phospholipid-dependent protein kinase (versus other protein kinases) the assay should be done in the presence of 0.5 mM $CaCl_2$ alone, 0.04 mg ml^{-1} PS alone, 0.5 mM $CaCl_2$ and 0.04 mg ml^{-1} PS together, and in the absence of both $CaCl_2$ and PS. Maximal activity for PKC should be obtained only in the presence of both Ca^{2+} and PS (Fig. 3).

PKC binds and is stimulated by diacylglycerols (DAG) (0.1–1.0 μg ml^{-1}) in addition to PS and lowers the free Ca^{2+} concentration necessary to obtain maximal stimulation (Kikkawa et al. 1983, Mori et al. 1982). However, the addition of diolein (0.1–1.0 μg ml^{-1}) to partially purified PKC at varying Ca^{2+} and PS levels shows little effect on the phosphorylation activity, which is most likely due to the presence of endogenous DAG. Similarly if the PS is not pure (i.e., if it contains hydrolysis products) you may see DAG contaminant stimulation. The purity of the PS is readily checked by TLC analysis (see above).

G. Binding of Phorbol Dibutyrate to Protein Kinase C

Equipment
- Sonicator
- Eppendorf centrifuge
- Liquid scintillation counter

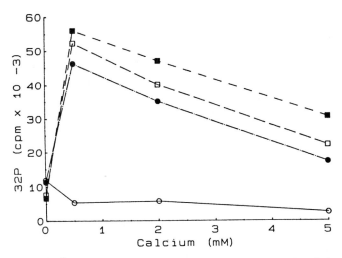

Fig. 3. Ca^{2+} and phospholipid stimulation of histone phosphorylation by partially purified PKC. Varying amounts of Ca^{2+} and phosphatidylserine were used in the above assay system. Partially purified PKC from bovine brain (Walsh et al. 1984) (maximal activity fraction from Phenyl Sepharose column, 20 μl/assay) was used as the enzyme source. *Open circles* absence of PS; *solid circles* 0.04 mg ml^{-1} PS; *open squares* 0.1 mg ml^{-1} PS; *solid squares* 0.25 mg ml^{-1} PS

Solutions needed
— Partial purified protein kinase C solution 2 mg protein ml^{-1}
— Tris/HCl-buffer   20 mM, pH 7.5
— Phosphatidylserine (PS)   1 mg ml^{-1}
        Preparation: 20 μl of a 10 mg ml^{-1} stock solution in Chloroform are
            put in a glass tube. The chloroform is removed with nitrogen. 200
            μl Tris-buffer are added and the PS is dispersed by sonication.
— CaCl$_2$ 100 mM
— Bovine IgG   4 mg ml^{-1} (Sigma)
— Phorbol-12,13-dibutyrate (PDB)   400 μM
— ^{3}H-PDB   160 nM
— ^{3}H-PDB   1.6 μM
— Polyethyleneglycol (PEG)   30% w/v in Tris-buffer
        MW 6000
— PPO/POPOP solution   25 g 2,5 Diphenyloxazole (PPO)
                                           1 g 2,2'-p-Phenylene-bis-(5-phenyl-
                                              oxazole) (POPOP)
                         disolved in 5 ml Toluene
— Scintillation cocktail   270 ml PPO/POPOP solution
                                           230 ml Triton X-100
                                           10.2 ml glacial acetic acid

For the binding assay the following mixtures are made in Eppendorf tubes
- Total binding   5 μl CaCl$_2$
-                                25 μl Phosphatidylserine
                                250 μl IgG
                                20 μl protein
                                x μl ^{3}H-PDB (see table below)
               make up to 400 μl total volume with Tris-buffer
- Nonspecific binding   5 μl CaCl$_2$
                                25 μl Phosphatidylserine
                                250 μl IgG
                                20 μl protein
                                x μl ^{3}H-PDB (see table below)
                                10 μl not radioactive PDB
               make up to 400 μl total volume with Tris-buffer

- ^{3}H-PDB concentrations	5	10	20	50	100 nM
- 160 nM ^{3}H-PDB added	12.5	25	50	–	– μl
- 1. 6 μM ” ”	–	–	–	12.5	25 μl

The samples are incubated for 30 min at room temperature and then for 15 min at 4°C. 370 μl PEG are then added to precipitate protein. The samples are allowed to stand 15 min at 4°C to complete precipitation and the precipitate is then pelleted by centrifugation in an Eppendorf centrifuge at 12,000 rpm for 15 min. The supernatant is discarded, the pellet redissolved in 1% Triton X-100, and after addition of 5 ml scintillation cocktail counted in a scintillation counter.

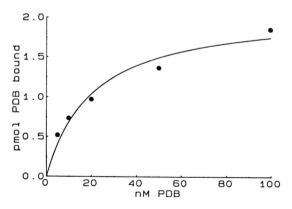

Fig. 4. Binding of ^{3}H-PDB to partially purified PKC

References

Babior BM, Cohen HJ (1981) Leucocyte function. Churcill Livingstone Ed, New York, pp 1−38

Bradford M (1976) Anal Biochem 72:248−255

Castagna M, Takai Y, Kaibuchi K, Sano K, Kikkawa U, Nishizuka Y (1982) J Biol Chem 257:7847−7851

Hecker E, Schmidt R (1974) Chem Org Naturst 31:377−467

Hecker E, Schmidt R (1979) Fortschr Chem Org Naturstoff 31:377−467

Inoue M, Kishimoto A, Takai Y, Nishizuka Y (1977) J Biol Chem 252:7610−7616

Kikkawa U, Minakuchi R, Takai Y, Nishizuka Y (1983) Methods Enzymol 99:288−298

Kuo JF, Andersson RGG, Wise BC, Mackerlova L, Salomonsson I, Brackett NL, Katoh N, Shoji M, Wrenn RW (1980) Proc Natl Acad Sci USA 77:7039−43

Mori T, Takai Y, Yu B, Takahasi J, Nishizuka Y (1982) J Biochem 91:427−431

Nishizuka Y (1984) Nature (London) 308:693−698

Schatzman RC, Raynor RL, Fritz RB, Kuo JF (1983) Biochem J 209:435−443

Sharkey NA, Blumberg PM (1985) Biochem Biophys Res Commun 133:1051−1056

Sharkey NA, Leach KL, Blumberg PM (1984) Proc Natl Acad Sci USA 81:607−610

Spector T (1978) Anal Biochem 86:142−146

Walsh MP, Valentine KA, Ngai PK, Carruthers CA, Hollenberg MD (1984) Biochem J 224:117−127

Extraction, Partial Purification, and Reconstitution of a Mixture of Carriers from the Inner Mitochondrial Membrane

MACIEJ J. NAŁĘCZ, KATARZYNA A. NAŁĘCZ and ANGELO AZZI

I. Aim of the Experiment

The present experiment is an example of procedures often employed in studies on the purification and characterization of membrane proteins. Solubilization of the inner mitochondrial membrane will be followed by the reconstitution of the solubilized proteins either directly or after a partial purification by hydroxylapatite chromatography. Subsequently the carrier-mediated accumulation of a substrate will be measured.

II. Introduction

The existence and the substrate/inhibitor specificity of mitochondrial carriers have first been inferred by using the swelling technique. Swelling occurs when large amounts of solutes enter the matrix space, causing increased osmotic pressure followed by water penetration and enlargement of the mitochondrial volume. The latter can be measured as a decrease of the light scattering of the mitochondrial suspension.

More recently, the nature and function of the mitochondrial carriers have been studied by directly observing the accumulation or the exchange of radioactive substrates. These studies afforded a clear demonstration of the existence of substrate carriers, a kinetic characterization of the different transport processes catalyzed by them and a clarification of their energetic requirements (for a review see LaNoue and Schoolwerth 1979).

Among them are the following:
- phosphate carrier (exchanges phosphate against OH$^-$)
- carnitine/acylcarnitine carrier
- ADP/ATP translocator
- oxoglutarate carrier (exchanges 2-oxoglutarate against malate)
- glutamate/aspartate exchanger
- dicarboxylate carrier (exchanges dicarboxylates against each other or with phosphate)

Membrane Proteins, ed. by Azzi
© Springer-Verlag Berlin Heidelberg 1986

— tricarboxylate carrier (exchanges citrate and isocitrate against phosphoenol-
pyruvate, malate and OH^-)

The mechanisms by which the transport occurs can be grouped into several
types (Mitchell 1979):
— proton-compensated electroneutral transport (phosphate, monocarboxy-
late, and glutamate carriers)
— electroneutral exchange (dicaroxylate, oxoglutarate, and tricarboxylate
carriers)
— electrogenic exchange (glutamate/aspartate and ADP/ATP exchangers)
— neutral metabolite transport in which protons are not involved (transport
of neutral amino acids and carnitine/acylcarnitine transport)

Recent studies have focused on the molecular mechanism of the transport,
together with the isolation of the carrier proteins. Some of the carriers from
the inner mitochondrial membrane have already been purified, e.g., the ATP/
ADP exchanger (Krämer and Klingenberg 1977) and the phosphate carrier
(Kolbe et al. 1981, 1984, Wohlrab et al. 1984). Other carriers have been extracted, partly purified, and reconstituted into liposomes, e.g., the tricarboxylate carrier (Stipani et al. 1980, Stipani and Palmieri 1983), the dicarboxylate
carrier (Saint-Macary and Foucher 1983, Kaplan and Pedersen 1985), the glu-
tamate/aspartate exchanger (Krämer 1984), the oxoglutarate carrier (Bisaccia
et al. 1985, M. Nałęcz et al. 1986) and the pyruvate carrier (K. Nałęcz et al.
1986).

III. Experimental Part

A. Materials and Buffers

Beef heart mitochondria are prepared by a standard procedure (Smith 1976).
Submitochondrial particles (inner mitochondrial membranes) are obtained
by sonication of the mitochondria and subsequent differential centrifugation
(Lee and Ernster 1967).
    The chemicals used are the following:
— Triton X-114, Triton X-100, EDTA (ethylenediaminetetraacetic acid
sodium salt), phenylsuccinate and Dowex 1-X8, Cl^- form (the Cl^- was
exchanged to formate before use) from Fluka, Buchs, Switzerland
— MOPS (morpholinopropanesulfonic acid), Folin reagent, PPO (2,5 Diphenyloxazole), POPOP (2,2 -p-Phenylene-bis (5-phenyloxazole) and malonic
acid from Merck.
— Radioactive malonate (1-^{14}C-malonate) from Amersham.

- Asolectin (extracted phospholipids from soybeans) from Associated Concentrates, Woodside, L. I., USA.
- Cardiolipin (5 mg ml^{-1} in methanol) from Sigma, St. Louis, MO, USA
- Hydroxylapatite (HTP-Biogel) from Bio-Rad, Richmond, CA, USA

All other reagents should be of analytical grade.

Buffer A for solubilization of the membranes and for hydroxylapatite chromatography: 4% Triton X-114, 50 mM NaCl, 1 mM EDTA, 20 mM MOPS and 2 mg ml^{-1} cardiolipin, pH 7.2. 6 ml of this solution have to be prepared freshly before the experiment. The solution is sonicated with cooling and under nitrogen until it is clear.

Buffer B to be enclosed in the liposomes: 50 mM NaCl, 10 mM malonate and 10 mM MOPS, pH 7.2. About 5 ml are needed for one experiment.

Buffer C for the transport measurements: 100 mM NaCl, 2 mM malonate and 20 mM MOPS, pH 7.2 and 5 μCi of radioactive malonate in 300 μl of buffer C.

170 mM sucrose (250 ml per experiment).

100 mM phenylsuccinate pH 7.2

Scintillation cocktail: 12.5 g PPO and 500 mg POPOP are dissolved in 2.5 l of toluene. 540 ml of this solution are mixed with 460 ml of Triton X-100 and 20.4 ml of glacial acetic acid.

Solutions for the protein determination
- Solution I: 2% Na$_2$CO$_3$, 0.3% SDS (Na-dodecylsulfate)
                  in 0.1 M NaOH
- Solution II: 2% sodium tartrate
- Solution III: 1% CuSO$_4$ · 5 H$_2$O

CuSO$_4$/NaTartrate/NaCO$_3$-mixture: 0.5 ml of solution II and 0.5 ml of solution III are supplemented with 49 ml of solution I. This mixture has to be prepared freshly before use.
- 20% SDS
- Folin reagent, diluted with water 1:1.5
- Bovine serum albumine 1 mg ml^{-1} in water

B. Equipment

- Ultracentrifuge
- Sonicator with microtip
- Liquid scintillation counter
- Thermostated water bath

— Vortex mixer
— Calibrated glass test tubes
— Normal glass test tubes
— Eppendorf tubes
— Small chromatography column (about 7 x 60 mm)
— Pasteur pipets
— Automatic pipets
— Scintillation vials
— Glass homogenizer.

C. Methods

All the procedures are performed at 4°C unless otherwise mentioned.

1. Solubilization

A 0.5-ml sample of freshly thawed submitochondrial particles (approximately 15 mg of protein) is mixed with 1.2 ml of buffer A and gently homogenized. The suspension is kept in ice for 20 min an then centrifuged at 100,000 g for 40 min. The brown-red pellet is discarded and the yellowish supernatant containing the solubilized membrane proteins is either directly used for reconstitution or first submitted to hydroxylapatite chromatography. An aliquot of the supernatant is kept for protein determination.

2. Hydroxylapatite Chromatography

600 mg of dry hydroxylapatite is placed into the chromatography column and loaded in the cold room with 0.6 ml of the supernatant. When all the supernatant has entered the gel, 1.8 ml of buffer A is loaded. Elution takes about 1 h. The eluate (about 0.6 ml) is used for reconstitution and for protein determination.

3. Reconstitution

Liposomes are prepared from asolectin by first swelling 480 mg of the dry lipid in 4 ml of buffer B for 2 h in the cold and subsequent sonication with cooling and under nitrogen until the suspension is opalescent. 360 μl portions of the liposomes are pipeted into Eppendorf tubes. 20 μl of either supernatant or hydroxylapatite eluate are added per portion of liposomes. The samples are mixed, allowed to stand on ice for 2 min, frozen in liquid nitrogen and subsequently left at room temperature for thawing. The viscous, white suspensions are now sonicated with cooling and under nitrogen until a transparent solution of vesicles is obtained (about 30 s). In order to remove the external malonate the vesicles are passed through Pasteur pipets filled with Dowex (1.5 ml of resin), equilibrated with 170 mM sucrose. After the 380-μl samples have entered the gel, the proteoliposomes are eluted with additional 600 μl of sucrose. Loading and elution can be accelerated by applying compressed air to the columns. From the moment of the appearance of the opalescent proteoliposomes at the bottom of the column, the eluate is collected into a calibrated tube until the columns are dry. Proteoliposomes of the same sort are collected into the same tube. Four samples will be reconstituted with the solubilized membrane material and six samples with the hydroxylapatite eluate, yielding about 3 ml and 4.5 ml of proteoliposomes, respectively. The final volume of the proteoliposomes is measured.

4. Transport (Malonate/Malonate Exchange)

300 μl aliquots of the respective proteoliposomes are used per sample. The samples are incubated for 1 min in a water bath at 30°C. Control samples are supplemented with 20 μl of 100 mM phenylsuccinate to inhibit the transport reaction. The reaction is started by the addition of 30 μl of radioactive substrate solution (buffer C), the final concentration of malonate outside being 0.2 mM. After the indicated time, the transport reaction is stopped with 20 μl of 100 mM phenylsuccinate and the samples are immediately processed further. To the "Zero-time" samples, a mixture of substrate and inhibitor is added. After the reaction the samples are passed quickly through Pasteur pipets as described above. The reaction mixtures (350 μl) are loaded on small Dowex columns and rapidly eluted with an additional 450 μl of 170 mM sucrose, squeezed through the Pasteur pipet using a rubber bulb. The eluates are collected into separate calibrated glass tubes from the time when the sample is applied to the column until the column is dry. The tubes are supplemented with 170 mM sucrose to a final volume of 2 ml. 200 μl aliquots of each sample are pipetted into scintillation vials, mixed with 5 ml of scintillation cocktail and counted for 2 min.

In the present experiment, the time course of the malonate/malonate exchange is measured after reconstitution of solubilized membranes (samples "S") and of the hydroxylapatite eluate (samples "E"), as indicated in the tables below. For each time point also a control sample with the inhibitor added from the beginning will be measured in order to distinguish the carrier-mediated exchange from the unspecific diffusion and binding of malonate to the proteoliposmes.

Further, samples of 5 or 10 μl of buffer C are also measured in the scintillation counter in order to determine the specific radioactivity of the substrate.

The following samples will be prepared:

Reconstituted supernatant

Sample	Inhibitor added at the beginning	Incubation time (s)
S	+	0
S	+	30
S	–	30
S	+	60
S	–	60
S	+	120
S	–	120
S	+	300
S	–	300

Reconstituted hydroxylapatite eluate

Sample	Inhibitor added at the beginning	Incubation time (s)
E	+	0
E	+	30
E	–	30
E	+	60
E	–	60
E	+	120
E	–	120
E	+	300
E	–	300
E	+	600
E	–	600

5. Protein Determination

The procedure of Lowry is modified here by the addition of SDS to avoid problems with high Triton X-114 and lipid concentrations. Bovine serum albumine (1 mg ml^{-1}) used as a protein standard is also dissolved in a buffer containing both Triton and lipid.

Standard curve

BSA (μl)	0	5	10	20	30	40
Water (μl)	100	95	90	80	70	60
Buffer A (μl)	100	100	100	100	100	100
20% SDS (μl)	50	50	50	50	50	50

Analysis

Supernatant (20 x diluted with buffer A) (μl)	100	—
Hydroxylapatite eluate (not diluted) (μl)	—	100
Water (μl)	100	100
20% SDS	50	50

To all samples 1 ml of the $CuSO_4$/Na-Tartrate/Na_2CO_3 is added, the samples are mixed, left at room temperature for 10 min and subsequently supplemented with 0.1 ml of the diluted Folin reagent. The absorption is measured after 30 min at 750 nm and the protein content of the samples is read from the standard curve.

IV. Results

The malonate/malonate exchange rate is calculated as nanomoles of malonate transported per mg of protein during a given time. Only the phenylsuccinate-sensitive transport is considered to be catalyzed by the carrier and should be taken into account. The latter is calculated as the difference between the transport samples and the control samples.

We first calculate the cpm of malonate transported per 20 μl of supernatant or hydroxylapatite eluate (one reconstituted sample). We have to take into account that out of a reconstituted sample of proteoliposomes (= total amount of proteoliposomes of one sort divided by the number of reconstitutions) we take an aliquot of 300 μl for the transport assay. Further, after transport the radioactivity is measured in aliquots of 200 μl out of a total volume of 2 ml.

This can be summarized in the following formula:

$$\frac{\text{cpm of malonate transported}}{20\ \mu l\ \text{reconstituted protein}} =$$

$$= \frac{\text{Volume of proteoliposomes}}{\text{Number of reconstitutions} \cdot 0.3 \text{ ml}} \cdot \frac{2 \text{ ml}}{0.2 \text{ ml}} \cdot \text{cpm}$$

It is assumed that under the conditions used all the proteoliposomes are eluted from the Dowex columns.

The cpm are converted into nmoles of malonate by using the radioactivity measurement of the aliquots of buffer C and the amount of protein in 20 μl of supernatant or hydroxylate eluate is calculated from the protein determination.

Typical results are summarized in the table below.

Sample	Control	Transport	Phenylsuccinate sensitive transport (transport minus control)
		(nmol malonate mg^{-1} protein)	
S 0 s	0.03	–	0.0
S 30 s	0.06	0.14	0.08
S 60 s	0.11	0.32	0.21
S 120 s	0.24	0.61	0.37
S 300 s	0.37	0.84	0.47
E 0 s	30.9	–	0.0
E 30 s	40.7	53.8	13.1
E 60 s	52.5	81.2	28.7
E 120 s	67.2	120.1	52.9
E 300 s	120.4	190.2	69.8
E 600 s	172.8	311.4	138.6

References

Bisaccia F, Indiveri C, Palmieri F (1985) Biochim Biophys Acta 810:362–369
Juliard JH, Gautheron DC (1973) FEBS Lett 37:10–16
Kaplan RS, Pedersen PL (1985) J Biol Chem 260:10293–10298
Kolbe HVJ, Böttrich J, Genchi G, Palmieri F, Kadenbach B (1981) FEBS Lett 124: 265–269
Kolbe HVJ, Costello D, Wong A, Lu RC, Wohlrab H (1984) J Biol Chem 259:9115–9120
Krämer R (1984) FEBS Lett 176:351–354
Krämer R, Klingenberg M (1977) FEBS Lett 82:363–367
LaNoue KF, Schoolwerth AC (1979) Annu Rev Biochem 48:871–922
Lee CP, Ernster L (1967) Methods Enzymol 10:543–550
Mitchell P (1979) Eur J Biochem 95:1–20
Nałęcz K, Bolli R, Wojtczak L, Azzi A (1986) Biochim Biophys Acta (in Press)

Nałęcz MJ, Nałęcz KA, Broger C, Bolli R, Wojtczak L, Azzi A (1986) FEBS Lett 196: 331–336
Saint-Macary M, Foucher B (1983) Biochem Biophys Res Commun 113:205–211
Smith AL (1976) Methods Enzymol 10:81–86
Stipani I, Palmieri F (1983) FEBS Lett 161:269–274
Stipani I, Krämer R, Palmieri F, Klingenberg M (1980) Biochem Biophys Res Commun 97:1206–1214
Wohlrab H, Collins A, Costello D (1984) Biochemistry 23:1057–1064

II. Lipid-Protein Interaction

The Effect of Polypeptides on the Fluidity of Lipid Bilayers

G. SARTOR, M.B. FERRARI and P. CAVATORTA

I. Introduction

The depolarization of fluorescence from hydrophobic fluorophores has been used to probe the fluidity or "microviscosity" within the hydrophobic region of membrane bilayer (Shinitzky et al. 1971).

These physical properties are of paramount importance to the understanding of the function of membranes, because any variation of them may result in an alteration of the physiological behavior of the membrane components. It is well established that several factors can influence the fluidity of bilayers, including cholesterol content, lipid composition, and proteins. In particular it has been previously reported (Hoffman et al. 1981) that the presence of protein influences the lipid behavior and that this can be monitored by using fluorescence polarization methods.

It has been pointed out by several authors (Kinosita et al. 1977) that the concept of "microviscosity" cannot be applied to describe the physical state of the membrane. The static fluorescence anisotropy may be considered in fact to be determined by two terms. One, the correlation time (ϕ), linked to the fluidity, and the other, the order parameter (S), related to the acyl chain spatial distribution or order of the membrane.

These two parameters can be evaluated from both static and time-resolved fluorescence anisotropy techniques. The purpose of this experiment is to test the effect of peptides on the fluidity of the membrane using a model system consisting of dimyristoyl-phosphatidylcholine (DMPC) vesicles with incorporated channels of the pentadecapeptide Gramicidin A (GA). The fluidity will be determined by measuring both static and dynamic fluorescence anisotropy of the probe 1,6-diphenylhexatriene (DPH).

Static Anisotropy

Static anisotropy measurements have been extensively used in detecting phenomena associated with changes of fluidity, although they are not sufficient per se to monitor the detailed structural behaviour of membranes. For example, DPH fluorescence anisotropy measurements have been shown to be a useful technique for detecting the transition temperature of several phospholipids and in monitoring changes due to the effect of external agents (Gomez-Fernandez et al. 1980). DPH is chosen as a fluorescent probe because it possesses nearly cylindrical symmetry, and consequently is thought to align with the hydrocarbon acyl chain of the lipids. Moreover, the absorption and emission dipole moments of DPH are nearly parallel (Zannoni et al. 1983). These properties simplify the mathematical analysis necessary to interpret the dynamic fluorescence observations.

II. Experimental Procedure

A. Chemical

— Chloroform (spectrofluorimetric grade)
— Methanol
— Tetrahydrofuran (spectrofluorimetric grade) (THF)
— Tris HCl buffer 20 mM pH 7
— Gramicidin A (from ICN Biochemicals, Cleveland, OH, USA)
— Dimyristoylphosphatidylcholine (from Avanti Lipids, Birmingham, AL, USA)
— 1,6-diphenylhexatriene (from Aldrich-Chemie, Steinheim, West Germany) 2 mM in THF

B. Equipment

— 15 ml Corex tubes
— Nitrogen or Argon gas cylinder
— Desicator
— Vacuum pump
— Vortexer
— Sonicator
— Bench centrifuge
— Spectrofluorimeter with temperature controlled cell holder
— Two polarizers

A sample of DMPC (15 mg) alone and another with DMPC (15 mg) and GA (2.5 mg) are dissolved in $CHCl_3:CH_3OH$ 2:1 (5 ml or more if necessary) solutions in 15 ml Corex tubes. The solvent is evaporated from each sample using a stream of nitrogen to obtain a thin film and the samples are left over-night under vacuum. After this, 5 ml of Tris HCl buffer is added and the samples are vigorously vortexed and sonicated for 30 min in steps of 6 min with intervals of 2 min, at 35°C (above the transition temperature). The suspension is then centrifuged for 30 min at 1500 g in order to sediment any large aggregates.

The concentration of the phospholipids is determined using the method of Marinetti (1962) and the concentration of GA is determined spectrophotometrically assuming an extinction coefficient of 22,000 mol^{-1} at 280 nm (Cavatorta et al, Part I, this Vol.). The samples are diluted in order to minimize the scattering (Cavatorta et al. Part I, this Vol.) and few microliters of DPH in TFH are added in order to achieve a phospholipid:DPH ratio of at least 300:1, the high ratio is necessary to avoid perturbation of the membrane (Shinitzky et al. 1971). The samples are finally kept for 30 min in the dark at 25°C to allow the probe to incorporate completely in the membrane.

The samples are excited using a polarizer inserted in the excitation light-path, oriented in vertical direction. The fluorescence is measured first with the analyzer in the emission lightpath, oriented vertically (I_{VV} or I parallel) and then with the analyzer oriented horizontally (I_{VH} or I perpendicular). The values of G, an instrumental correction factor should be measured on one of the samples using horizontally polarized excitation light and vertical and horizontal emission intensities (Dale et al. 1977).

$$G = I_{HV}/I_{HH}.$$

The anisotropy is then calculated at several temperatures as:

$$r = (I_{VV} - I_{VH}G) / (I_{VV} + 2I_{VH}G).$$

The anisotropy values are plotted vs. temperature and the effect of GA on the fluidity of the lipids may be determined. What we expect is an increase of anisotropy in the sample containing the GA with respect to DMPC alone, the increase being the result of the rigidifying effect of the polypeptide on the lipid chains.

A typical result of this experiment is shown in Fig. 1.

C. Time-Resolved Anisotropy Decay

As previously pointed out, the static anisotropy gives only a time-averaged picture of the dynamics of the membrane. Anisotropy decay experiments are needed in order to distinguish between static (order of the acyl chains) and dynamics (fluctuation) of the fluidity (Zannoni et al. 1983).

An anisotropy decay experiment should allow us to see if the effect of GA on acyl chains is either static or dynamic or both.

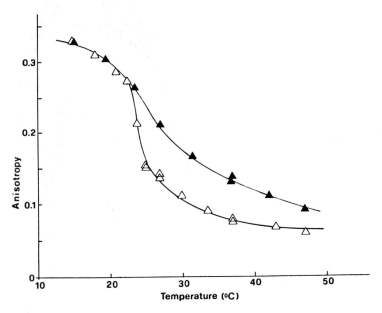

Fig. 1. Dependence of the static fluorescence anisotropy of DPH in DMPC vesicle with GA (*filled symbols*) and without GA (*open symbols*)

III. Experimental Procedure

A. Chemicals

Same as the static anisotropy experiment.

B. Equipment

Single photon counting-fluorescence apparatus equipped with:
— two polarizers
— temperature-controlled cell holder

The anisotropy of DPH in DMPC in absence and in presence of GA is evaluated at each time as:

$$r_{(t)} = (I_{VV(t)} - I_{VH(t)}\, G)/(I_{VV(t)} + 2I_{VH(t)}G).$$

The decay with time of $I_{VV(t)}$ and $I_{VH(t)}$ is separately obtained counting for the same period of time (normally 2000 s). The fluorescence photons of the sample with the excitation and emission polarizer both vertical $[I_{VV(t)}]$

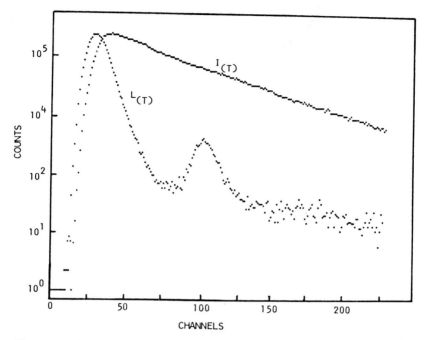

Fig. 2. Total fluorescence decay of DPH $[I_{(t)}]$ and lamp emission profile $[L_{(t)}]$. Channel width is 0.178 ns. The peak around 18 ns is an after pulse peak characteristic of the photomultiplier (O'Connor et al. 1984)

is first measured and then, with emission polarizer turned horizontally, is measured $I_{VH(t)}$.

The sum:

$$I_{(t)} = I_{VV(t)} + 2I_{VH(t)}G$$

represents the total intensity decay of DPH in membrane, as shown in Fig. 2, and it follows the relation:

$$I_{(t)} = \sum_i a_i\, e^{-t/\tau_i},$$

where a_i are pre-exponential factors and τ_i are the fluorescence lifetimes of DPH.

Due to the cylindrical symmetry of the probe the fluorescence anisotropy is assumed to decay with time according to the following relation:

$$r_{(t)} = (r_0 - r_\infty)\, e^{-t/\phi} + r_\infty,$$

where ϕ is the correlation time, expressed in nanoseconds, r_0 is the anisotropy at zero time. r_∞ is the infinite anisotropy related to the order parameter of the

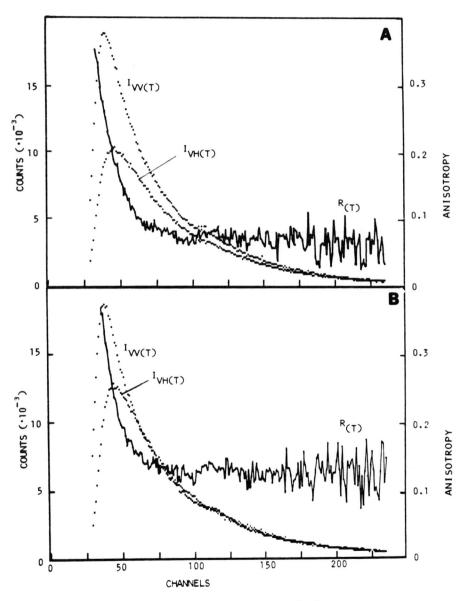

Fig. 3. Parallel (I_{VV}), perpendicular (IV_H) and anisotropy $[r_{(t)}]$ profiles in DMPC (**A**) and in DMPC + GA (**B**). Channel width is 0.178 ns

system by the following equation:

$$S = (r_\infty/r_0)^{1/2}.$$

Becaused the finite width of the flashlamp emission profile, the experimental decay of $I_{VV}(t)$ and $I_{VH}(t)$ is not a true decay but is a convolution of the lamp profile and sample intensity decay. Being the fluorescence lifetime of the probe in the nanoseconds time scale, it must be deconvoluted in order to obtain the anisotropy function.

The deconvolution and the evaluation of ϕ and r are carried out together through a nonlinear least-squares analysis program (Dale et al. 1977) using a lamp profile obtained by means a scattering substance (glycogen). A typical anisotropy decay profile is shown in Fig. 3.

References

Dale RE, Chen LA, Brand L (1977) J Biol Chem 252:7500−7510

Gomez-Fernandez JC, Goni FM, Bach D, Restall CJ, Chapman D (1980) Biochim Biophys Acta 598:502−516

Hoffman W, Pink DA, Restall CJ, Chapman D (1981) Eur J Biochem 114:585−589

Kinosita K, Kawato S, Ikegami A (1977) Biophys J 20:289−305

Lakowicz JR (1983) Principles of fluorescence spectroscopy. Plenum Press, New York London

Marinetti GV (1962) J Lipid Res 3:1−8

O'Connor DV, Phillips D (1984) Time correlated single photon counting. Academic Press,, London New York

Shinitzky M, Dianoux AC, Gitler C, Weber G (1971) Biochemistry 105:2106−2113

Zannoni G, Arcioni A, Cavatorta P (1983) Chem Phys Lipids 31:179−185

Polypeptide-Lipid Interactions as Studied by ^{13}C NMR

A. SPISNI, G. FARRUGGIA and L. FRANZONI

I. Introduction

The Gramicidin A-Lysolecithin model

There is much interest in the study of interactions between lipid bilayers and proteins or polypeptides as they reciprocally modulate their structure and function (Burnell et al. 1981, de Krujiff and Cullis 1980, Pink et al. 1981, Spisni et al. 1979, Tamm and Seelig 1983). Gramicidin A (GA), due to its highly hydrophobic character, is a polypentadecapeptide able to incorporate in lipid bilayers forming helical structures (Urry et al. 1979a,b, Wallace et al. 1981, Weinstein et al. 1980) and therefore it constitutes a very good model to study the behavior of the hydrophobic part of proteins. Lysolecithin (LY), as is well known, is a phospholipid that does not form bilayers in water but dissolves, forming micelles (Junger et al. 1970, Saunders 1966). However, when GA is allowed to interact with LY micelles for several hours at high temperature, it is able to force the LY molecules to assume a bilayer organization in which the polypeptide is incorporated, interacting strongly with the fatty acid chains (Pasquali-Ronchetti et al. 1983; Spisni et al. 1983).

II. Basic Principles of Nuclear Magnetic Resonance (NMR)

This section represents a very short description of the origin and behavior of an NMR signal. For more details see James (1975) and Slichter (1980).

A. The NMR Signal

At the basis of the NMR phenomenon lies the idea that atoms possess an angular momentum or spin that is normally indicated with $\vec{I}$. This means that we are dealing with small charges spinning on their axis and therefore creating

Membrane Proteins, ed. by Azzi
©Springer-Verlag Berlin Heidelberg 1986

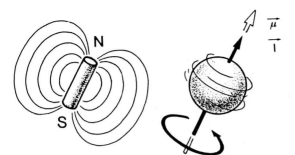

Fig. 1. A nucleus with an angular momentum $\vec{I}$ different from zero behaves like a microscopic magnetic bar. The strength and the direction of the magnetic field produced by its spinning is expressed by the magnetic moment $\vec{\mu}$

a magnetic field to which is associated a magnetic moment, $\vec{\mu}$. The intensity and the direction of the vector represent the strength and the direction of the field itself. Thus we can compare the behavior of an atom to a small magnetic bar as shown in Fig. 1.

It is also worthwhile to point out that $\vec{\mu}$ and $\vec{I}$ are related one to the other by Eq. (1):

$$\vec{\mu} = \gamma \vec{I} ,\qquad(1)$$

where γ is a constant characteristic of the nucleus and is called the gyromagnetic ratio.

Figure 2a,b shows the two conditions in which the atoms will be in the absence or presence of an external magnetic field B_0. In the simplest case of nuclei with spin $I=1/2$, as for ^{13}C, when there is no external field all the nuclei will be randomly oriented with respect to each other (Fig. 2a). On the contrary, if an external magnetic field B_0 is applied, we will have two possible states: one in which the nuclear spins are parallel and another in which they are antiparallel to the direction of B_0 (Fig. 2b).

These two orientations are characterized by two distinct energies, and it can be easily demonstrated that the energy E of each state is:

$$E = \mu_z B_0 ,\qquad(2)$$

knowing that:

$$\mu_z = (h/2\pi)\,\gamma\,I ,\qquad(3)$$

therefore the energy difference, ΔE, between the two spin states is:

$$\Delta E = h\nu = |\,(h/2\pi)\,\gamma\,B_0\,\Delta I| ,\qquad(4)$$

as for nuclei with spin $I=1/2$, $\Delta I=1$, Eq. 4 becomes:

$$h\nu = |\,(h/2\pi)\,\gamma\,|\,B_0 .\qquad(5)$$

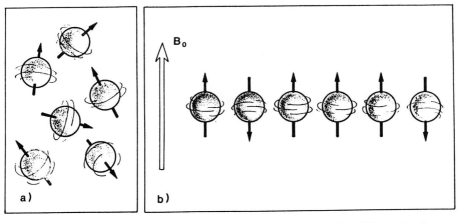

Fig. 2. a In the absence of an external magnetic field, the nuclear spins will be randomly oriented with respect to each other. **b** When an external magnetic field B_0 is applied to the nuclear system, if the nuclei possess an angular momentum I= 1/2, they will orient their magnetic moment vector $\vec{\mu}$, or spin, in two possible ways, either parallel or anti-parallel to the external magnetic field direction

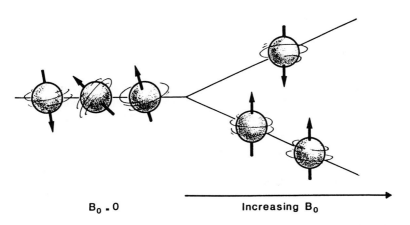

Fig. 3. The nuclei that have the spin parallel to the direction of the external magnetic field B_0 possess a slightly lower energy than those that are antiparallel. The energy difference between these two spin states is directly proportional to the strength of the external magnetic field: $h \nu = |(h/2\pi) \gamma| B_0$

The relevance of Eq. (4) is to show that the increase in intensity of the external field B_0 produces an increase of the separation of the two energy levels as shown in Fig. 3.

Another effect due to the presence of an external magnetic field B_0, over the spin system, is to induce a synchronous motion of the nuclear spins around the B_0 axis. The angular velocity ω_0 associated to this motion is called the Larmor frequency and is directly proportional to the external magnetic field. Figure 4a depicts the precession of the nuclear spins.

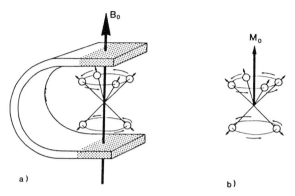

a)   b)

Fig. 4. a Besides inducing an alignement of the nuclear magnetic moments, the interaction of the external magnetic field B_0 with the nuclear spins creates a torque that forces the nuclei to process around the B_0 axis at an angular velocity ω_0, called Larmor frequency. This angular velocity is directly related to the strength of the field by the equation: $\omega_0 = \gamma B_0$. **b** The alignment of the nuclear spins and the slightly higher number of nuclei parallel to the external magnetic field B_0 are enough to produce a net macroscopic magnetization, M_0, pointing in the same direction of the external field

At thermal equilibrium, according to the Boltzman distribution, the population of the lower level will be slightly higher than that of the upper level, so that the ratio between them will be: $\exp(\Delta E/KT)$. The slightly higher number of spins in the lower level, however, is enough to give rise to the macroscopic magnetization, M_0, (Fig. 4b), the magnitude of which is a function of the external magnetic field, B_0, of the number of spins, N, of the angular momentum, $I\hbar$, of the gyromagnetic ratio, γ and of the temperature, T, according to Curie's Law:

$$M_0 = \frac{N \gamma^2 \hbar^2 I(I+1)}{3kT} B_0 \quad . \tag{6}$$

Thus, upon irradiation of the system with electromagnetic waves possessing an energy equal to the ΔE between the two spin levels or, in other terms, vibrating at a frequency equal to the Larmor frequency, ω_0, of the nucleus studied, it will be possible to induce a net absorption corresponding to a transition of the nuclear spins from the lower level to the upper level. In a normal NMR experiment the energy necessary to induce such a transition is in the range of the radiofrequency (rf), that is to say between 10 and 600 MHz, depending upon the field used and the nucleus under investigation.

As soon as the rf is switched off, the nuclear system will begin to relax back to thermal equilibrium and the absorbed energy will be released as a rf with the same frequency and intensity of the one absorbed: this detectable rf is the NMR signal.

B. The Relaxation Mechanism

As has been previously described, when the nuclei are irradiated with the proper rf, the Boltzman population distribution between the two levels is perturbed. The process by which they return exponentially to thermal equilibrium is called *relaxation.*

In principle, a nuclear spin can relax from a high energy state to a low energy state via spontaneous emission and/or stimulated emission. Spontaneous emission is a case in which the nuclear spin spontaneously jumps to a lower state while emitting a photon. However, since the probability for spontaneous emission depends on the third power of the frequency, being in the radio frequency range, this term is too small to be significant. This means that all NMR transitions are stimulated. There are several mechanisms that govern the relaxation process; however, for nuclei with spin 1/2, all of them have one thing in common: they are due to the interactions of the nuclear magnetic moments with a surrounding fluctuating magnetic field. In fact, as the only property of a nucleus with spin 1/2 which depends on the orientation is its magnetic moment, transitions between the spin levels can only be induced by magnetic fields. It is important also to remember that for the interaction to be effective, the time-varying magnetic field must fluctuate at the Larmor frequency of the nucleus under investigation. In fact, this interaction is of the same kind as the one utilized to induce an absorption transition.

There are a variety of possible sources for the fluctuating magnetic field, each of which causes a specific "spin-lattice relaxation" mechanism. Let us assume, for instance, that a molecule possesses an unpaired electron which we know has a magnetic moment about 1000 times larger than a proton and even more with respect to a ^{13}C nucleus. If the molecule tumbles randomly, a fluctuating magnetic field will be generated by that electron at the site of the nucleus. Then the component of that randomly varying magnetic field with a frequency that equals the Larmor frequency of the nucleus under investigation will be effective in the stimulation of transitions between the spin levels. This sort of relaxation mechanism governed by a magnetic field generated by an unpaired electron is called *paramagnetic relaxation.* In other cases the fluctuating magnetic field is generated by the nuclear magnetic moment of another nucleus and is called *nuclear dipole-dipole relaxation.* For nuclei with spin 1/2, this is the most common of the relaxation mechanisms.

Therefore, the important point that has to be realized is that as the nuclear spin-lattice relaxation process is usually dependent upon the molecular motions generating the randomly varying magnetic field, valuable information about these motions can be obtained by the study of the relaxation rates. The spin-lattice relaxation time is indicated with the symbol T_1 and it is also called longitudinal relaxation as it describes the relaxation of the longitudinal component (usually the z axis component) of the macroscopic magnetization M_0.

Besides the spin-lattice relaxation, there are several other relaxation processes, the most relevant of which is the *spin-spin relaxation* or transverse relaxation, characterized by the relaxation time T_2 and that refers to the relaxation of the components of the magnetization M_0 precessing on the x-y plane. T_2 is inversely proportional to the linewidth accordung to the relation:

$$\Delta\nu_{1/2} = \frac{1}{\pi\,T_2}\,, \tag{7}$$

where $\Delta\nu_{1/2}$ is the resonance linewidth at half height. However, for fast molecular motions, when the extreme narrowing conditions are met, $T_1 = T_2$, and therefore T_2, does not offer any relevant additional information. For further discussion on these and other relaxation mechanisms the reader is referred to references (James 1975, Slichter 1980).

It is possible to demonstrate that:

$$T_1^{-1} \simeq \frac{2\,\tau_c}{1 + \omega_0{}^2\,\tau_c{}^2}\,, \tag{8}$$

where ω_0 is the Larmor frequency in radians and τ_c is the correlation time for the particular motion considered. For a molecule randomly tumbling, τ_c measures the average time it requires to progress of one radian. In the extreme narrowing conditions, where $\omega_0{}^2\,\tau_c{}^2$ 1 and that for a ω_0 in the Mrad s^{-1} range corresponds to τ_c in the range of 10–100 ps, then Eq. 8 reduces to:

$$T_1^{-1} \propto \tau_c\,, \tag{9}$$

implying that the spin-lattice relaxation rate increases when τ_c increases, i.e., as the mobility of the molecule decreases. Thus, modifications of the molecular mobility should be reflected by changes of the relaxation time producing variation in the amplitude and intensity of the resonance line.

In the specific case of ^{13}C NMR, given the fact that the spin-lattice relaxation is mainly dependent on dipole-dipole interactions and that we are in the extreme narrowing conditions, it is possible to interpret the broadening and the associated decrease of intensity of the resonances as due to a decrease in the mobility of the nuclei under investigation.

III. Aim of the Experiment

The aim of the following experiment is to demonstrate, monitoring the ^{13}C
NMR signals of the lipid phase, that indeed the interaction between the peptide molecules and the phospholipids takes place only at the fatty acid site.
In fact, the strong hydrophobic interactions between the polypeptide residues
and the fatty acid side chains should reduce their molecular motion, and therefore a broadening and a consequent decrease of the resonance lines' intensity
is expected. On the other hand, the spectral lines from the polar part of the
phospholipids, being unaffected by those interactions, should not present any
relevant modification.

IV. Experimental Procedures

A. Reagents

— Lysolecithin (LY)
— Gramicidin A (GA)
— NaCl 100 mM
— D_2O 98%

B. Equipment

— NMR Spectrometer
— Sonifier equipped with microtip
— Thermostatable bath (25—100ºC)
— Vortex mixer
— VSL pyrex tubes with cap
— 10 mm NMR precision tubes

C. Preparation of the Samples

1. Reference Sample: LY Micelles

In a VSL pyrex tube add about 40 mg of LY to 2 ml of D_2O. Vortex the
sample until the LY is completely dissolved (the solution must be perfectly
clear). Add NaCl to a final concentration of 10 mM.

2. GA/LY Micellar Dispersion

In a VSL pyrex tube add about 50 mg of LY to 2 ml of D_2O and dissolve the lipid completely by vortexing for 1–2 min. Weigh out the required amount of GA to obtain a polypeptide: lipid molar ratio of 1:10 (MW GA = 1882; MW LY = 508). Add the GA to the lipid solution in small amounts and after each addition vortex the sample carefully until the GA is completely dispersed. After all the GA has been added, sonicate the sample for about 3 min using a sonifier with a microtip. Finally add NaCl 100 mM to a final concentration of 10 mM.

3. GA/LY Membrane System

Repeat the preparation of step (B), changing only the final concentration of the NaCl, that now will be 5 mM. Once the sample is ready, wrap it in aluminum foil and place it in a thermostatable bath at 80°C for 7–8 h.

D. NMR Measurements

Transfer each of the samples in a 10-mm NMR precision tube and set the spectrometer frequency at the value for ^{13}C, that at a field of 4.7 T is 50.33 MHz, and run the three spectra. Usually at this field strength a sweep width of 12000 Hz is used. The spectra are proton decoupled and an interval of 1.5s after each acquisition is generally long enough to allow for a complete relaxation of the ^{13}C nuclei. An exception might be the carbonyl resonance, which, having a longer relaxation time, requires a much longer pulse interval. Compare the spectra of samples (2) and (3) with the reference (1) and determine in which of the two samples GA is interacting with the lipids' fatty acid side chains, suggesting therefore the formation of bilayers in which the polypeptide is incorporated.

Acknowledgment. The authors gratefully acknowledge for the valuable help of Miss Edvige Masini in preparing the illustrations.

References

Burnell L, Alphen A van, Verkleji A, Krujiff B de (1980) Biochim Biophys Acta 597: 492–501
Burnell L, Clark ME, Hinke JA, Chapman NR (1981) Biophys J 33:1–26
James TL (1975) Nuclear magnetic resonance in biochemistry. Academic Press, London New York

Junger M, Hahn MH, Reinhauer H (1970) Biochim Biophys Acta 211:381–388

Krujiff B de, Cullis PR (1980) Biochim Biophys Acta 602:477–490

Pasquali-Ronchetti I, Spisni A, Casali E, Masotti L, Urry DW (1983) Biosci Rep 3:127–133

Pink DA, Georgallis A, Chapman D (1981) Biochemistry 20:7152–7157

Saunders L (1966) Biochim Biophys Acta 125:70–74

Slichter CP (1980) Principles of magnetic resonance. Springer, Berlin Heidelberg New York

Spisni A, Khaled MA, Urry DW (1979) FEBS Lett 102:321–324

Spisni A, Pasquali-Ronchetti I, Casali E, Lindner L, Cavatorta P, Masotti L, Urry DW (1983) Biochim Biophys Acta 732:58–68

Tamm LK, Seelig J (1983) Biochemistry 22:1474–1483

Urry DW, Spisni A, Khaled MA, Long MM, Masotti L (1979a) Int J Quant Chem Quant Biol Symp 6:289–333

Urry DW, Spisni A, Khaled MA (1979b) Biochem Biophys Res Commun 88:940–949

Wallace BA, Veatch WR, Blout WR (1981) Biochemistry 20:5754–5760

Weinstein S, Wallace BA, Morrow JS, Veatch WR (1980) J Mol Biol 143:1–19

Conformational Changes in Polypeptides and Proteins Brought About by Interactions with Lipids

L. MASOTTI, J. VON BERGER, and N. GESMUNDO

I. Introduction: Circular Dichroism and Optical Rotation

One important feature of biopolymers belonging to different classes such as proteins, nucleic acids, and polysaccharides is the helical structure. The recognition, quantitative determination and monitoring of the changes of such structures is of great importance in understanding the relationships between structure and function of biological macromolecules.

Optical rotation measurements represent an important tool for structural studies on biomolecules in solution: in fact they are based on the asymmetry of structure and on the response to polarized light of molecules.

When using optical rotation in the study of optically active molecules, one may employ plane polarized light (Fig. 1a).

As is well known, an electromagnetic wave is characterized by the amplitude and orientation of the electric (E) and magnetic (H) vectors that will always be perpendicular to one another and perpendicular to the direction of propagation.

To an observer looking in the negative X direction, E is seen to oscillate in the XZ plane. Thus the linearly polarized light can be also described as a combination of left circularly polarized light (E_L) and right circularly polarized light (E_R) of equal amplitude and traveling in phase.

This is depicted in Fig. 1b, where one looks directly toward the source, and the vector (E_z) sum of E_L and E_R will oscillate along the dotted line.

An example of circularly polarized light is shown in Fig. 2a, where the left component is represented.

It can be described, as represented in Fig. 2b, as a combination of two linearly polarized waves, whose electric field vectors oscillate in plane perpendicular to one another (in the figure E_1 is in the XZ plane and E_2 is in the XY plane), and out of phase of 90°.

Membrane Proteins, ed. by Azzi
©Springer-Verlag Berlin Heidelberg 1986

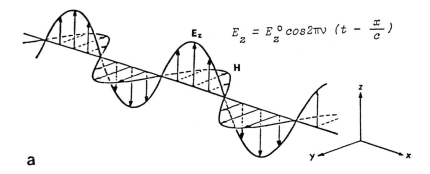

$$E_z = E_z^o \, cos2\pi\nu \left(t - \frac{x}{c}\right)$$

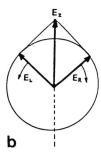

Fig. 1. a Plane (linearly) polarized radiation. **b** Plane polarized light as the resultant of the two circularly polarized vectors, E_L and E_R

II. Optical Activity

In order to be optically active a molecule must possess neither a plane nor a center of symmetry. As a consequence it will interact differently with left and right circularly polarized light. This differential interaction results in two different but related phenomena, i.e., optical rotatory disperson (ORD) and circular dichroism (CD).

As seen previously, light can be described as an oscillating electric vector which, when it interacts with a molecule, effects a momentary change in distribution of the charged particles that comprise the molecule, i.e., positive nuclei and negative electrons, and therefore changes the dipole moment of the molecule. As the molecule relaxes, the dipole moment is released to the beam. If the molecule contains a plane or a center of symmetry, it interacts equally with E_L and E_R. If the molecule is dissymmetric, then a preferential, stronger interaction with one of the circularly polarized components will take place. Such a component will have a lower velocity in the medium with a resulting larger refractive index. If, for example, E_L interacts more strongly with the molecule, then we shall have:

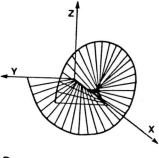

a

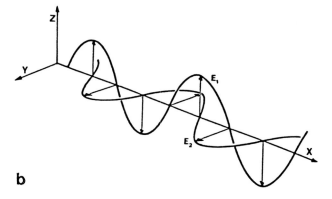

b

Fig. 2. a Left circularly polarized light. **b** The same circularly polarized light represented as the sum of the two electric vectors E_1 and E_2, perpendicular to one another and 90° out of phase

$$n_L = \frac{c}{v_L} \quad > \quad n_R = \frac{c}{v_R} \quad .$$

This phenomenon is termed circular birefringence. In any case, if $n_L \neq n_R$, the rotation velocity of E_L and E_R in the sample will be different; as a consequence the two components will have become out of phase and they will finally travel again with the same velocity on emerging from the sample.

The resultant of E_L and E_R, still plane polarized, emerging from the sample, (Fig. 3a) will be rotated at a certain angle with respect to the entering beam. The measurement of a results, when plotted as a function of wavelength, in an ORD curve.

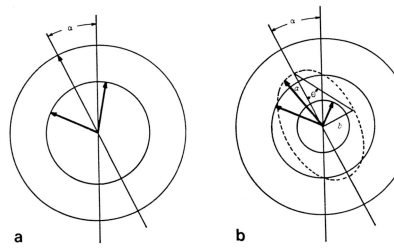

Fig. 3. a Circular birifrangence.

$$a' = \frac{\pi}{\lambda}(n_L - n_R),$$

where n_L and n_R are the refractive indices of the left and right component respectively.
b Circular dichroism.

$$\theta' = \frac{\pi}{\lambda}(k_L - k_R),$$

where k_L and k_R are the extinction coefficient of the left and right beam respectively.
The ellipticity is defined by the relationship:

$$\tan^{-1}\theta = \frac{b}{a}$$

In an absorption region, there will be a differential absorption of the two circularly polarized components, $(\epsilon_L \neq \epsilon_R)$ and therefore a dichroism (Fig. 3b).

The two components will have unequal amplitude and their contribution will result in an elliptically polarized light, characterized by both an ellipticity (the ratio of axes of the ellipse) and the direction of the major axis. The difference absorbance of circularly polarized beams results, when plotted as a function of wavelengths, in a CD curve.

In CD one observes positive or negative bands ($\epsilon_L > \epsilon_R$ and $\epsilon_L < \epsilon_R$ respectively) which approximately correspond to absorption bands (see also Fig. 5).

III. Parameters for Measuring Optical Activity

ORD. One way of reporting the rotation of linearly polarized light for a substance is in terms of *specific rotation*, defined as:

$$[a]_\lambda^t = \frac{a}{\rho 1} ,$$

where a is given in degrees, λ is most commonly reported for the sodium D line, t, the temperature is noted, ρ is the concentration in g cm^{-3} and 1, the pathlength, in decimeters. The *molar rotation* $[M]_\lambda$, is defined

$$[M]_\lambda = \frac{a}{\rho 1} \frac{MW}{100} ,$$

where MW is the molecular weight. The molar rotation can be expressed in terms of molar concentration C, and path length in cm as

$$[M]_\lambda = \frac{100a}{C 1} .$$

Finally a commonly used parameter for macromolecules is the *mean residue rotation* $[m]_\lambda$, defined as

$$[m]_\lambda = \frac{a}{\rho 1} \frac{MRW}{100} ,$$

where MRW is the mean residue weight of the monomers comprising the macromolecule.

CD. In circular dichroism the observable measured is a *difference absorbance,* i.e., ΔA, which is defined as

$$\Delta A = \Delta \epsilon \, cl = A_L - A_R ,$$

where A_L and A_R are the absorbances of the left and right circularly polarized beams, $\Delta \epsilon$ is $\epsilon_L - \epsilon_R$.

Commonly used is the molar ellipticity $[\theta]_\lambda$, related to ΔA by the relationship

$$[\theta]_\lambda = \frac{3300 (A_L - A_R)}{c 1} ,$$

where c is the concentration in mol^{-1} and 1 the pathlength in cm.

An equivalent espression is:

$$[\theta]_\lambda = 3300 (\epsilon_L - \epsilon_R) \ (\text{degree cm}^2 \text{ dmol}^{-1}) .$$

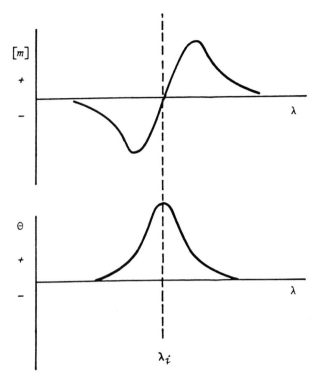

Fig. 4. The Cotton effect arising from the absorption of light by an optically active mole-cule. The wavelength dependence of the optical rotation is shown above and of the ellipticity below

IV. The Cotton Effect. Relationship Between Absorbance, CD and ORD

The relationship between absorbance and optical rotation is illustrated in an idealized form in Fig. 4. The upper curve represents the changes of optical activity with wavelength near and through the absorption band, or dichroic band, which is represented by the lower curve. At a longer wavelength the curve becomes more and more positive, exhibits a maximum, then becomes less positive, reaches zero and then becomes negative at a shorter wavelength. The phenomenon of passage of ORD through a maximum (or minimum), zero, and a minimum (or maximum) is called the Cotton effect. In the case illustrated in Fig. 4 it is a positive Cotton effect; the opposite behavior would be a negative one.

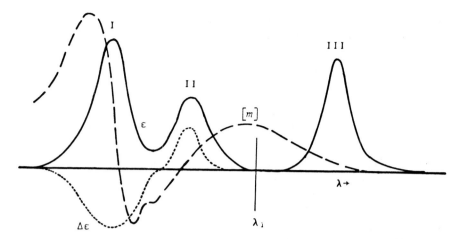

Fig. 5. The relationship of absorbance (ϵ), CD ($\Delta\epsilon$) and ORD ([m]) for a hypothetical molecule

Also CD changes with wavelength near and through the absorption band (Fig. 5). The *maximum* of the CD curve is positioned at the same wavelength of the absorption curve, but while absorption is always positive, CD can be either positive or negative.

Furthermore, in ORD the *position* of the Cotton effect lies at the λ where [m] is zero, while in CD it is positioned at the wavelength when there is a positive or negative CD maximum.

Figure 5 reports the absorbance (ϵ) CD ($\Delta\epsilon$) and ORD ([m]) for an idealized molecule and shows advantages and disadvantages of the techniques. In CD measurements each band corresponds to an absorption band and in Fig. 5 the negative CD of band I is well separated from the positive CD of band II. More resolution is then available in this case, while the ORD curve is spread over a wider range of wavelengths. At higher than λ_1 though, ORD would still give us some information, while CD, in the case of band III that is optically inactive, would not.

V. The Physical Basis of Optical Activity and Optically Active Chromophores

When a molecule absorbs light, a transition from the ground state to the excited state occurs involving a displacement of charge. This change in distribution of charges is called the transition dipole moment, denoted by μ. The transition can also have a circular component that corresponds to a rotating current. To the latter is associated a magnetic dipole moment, m, perpendicular to the plane of the circular motion. If μ and m each have a finite value, optical activity ensues. This is to say that a helical displacement of charge takes place and the left and the right circularly polarized waves will interact differently with the chromophore.

The analysis of an optically active molecule can be carried out by identifying moieties which are relatively independent of the molecule in which they are found. Such moieties are, for example, the methyl moiety, the peptide moiety, the disulfide moiety, etc.

Their interaction with light further classifies them as chromophoric and nonchromophoric, the latter being those that absorb light only in the vacuum ultraviolet (e.g., the methyl moiety).

A chromophoric moiety within a molecule can be inherently symmetric or inherently asymmetric. In the first case, the chromophore contains a plane or a center of symmetry. Its optical activity will then arise from its interaction with dissymmetrically located neighboring moieties (for example, a peptide moiety is symmetric and derives its chirality from interactions with other peptide moieties, methyl group etc.). They are further characterized either by weak absorptions (e.g., $n-\pi^*$ transition), or by strong absorption (e.g., $\pi-\pi^*$ transitions). A dissymmetric moiety is optically active regardless of its location within the molecule and constitutes one source of rotational strength (e.g., the disulfide bridge).

CD is widely used to determine the secondary structure of polypeptides and proteins. Since this is the aim of the experiment, the CD spectra of the α-helix, β-sheet, and the random coil conformations of the model polypeptide L-lysine are shown in Fig. 6.

VI. Experimental Part

A. Introduction

Gramicidin A (GA) is a polypentadecapeptide polymorphic in nature as its conformation is dependent on concentration, temperature, and solvent system (Urry et al. 1975).

It has been demonstrated that it can incorporate in membranes forming transmembrane channels that are specific for monovalent cations (Hladky and Haydon 1972). Moreover, the channels have been proposed to be formed by N-terminal to N-terminal dimers (Szabo and Urry 1979) bridged by

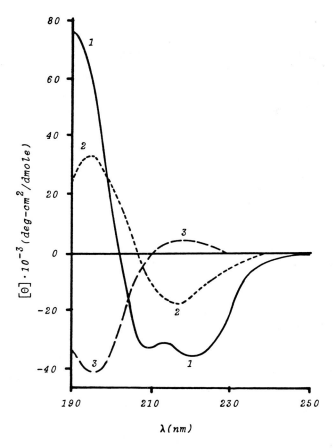

Fig. 6. CD of the model polypeptide L-lysine in *1* α-helix; *2* β-sheet; *3* random coil conformations. (After Greenfield and Fasman G 1969 Biochemistry 8:4118)

hydrogen bonds and the conformation of the polypeptide in this state has been suggested to be a single-stranded left-handed β-helix (Urry et al. 1967; Urry 1971; Urry et al. 1979).

There is experimental evidence that the incorporation of GA in lysolecithin disperson requires the onset of hydrophobic interactions between the polypeptide residues and the phospholipid fatty acid chains, and besides inducing a drastic modification of the lipid organization, produces dramatic changes of the polypeptide conformation (Masotti et al. 1979; Urry et al. 1967; Urry et al. 1979).

B. Aim of the Experiment

The aim of this experiment is to follow the conformational changes that Gramicidin A undergoes during incorporation in a lysolecithin dispersion, by using the circular dichroism technique. The results obtained are then related to the experiments carried out on the same step with fluorescence and NMR.

C. Experimental Procedures

1. Reagents

— Lysolecithin
— Gramicidin A
— NaCl 100 mM

2. Instruments

— Dicrograph (Jasco J 500A)
— Cylindrical cuvettes of 0.2 mm pathlength
— Vortex mixer
— VSL pyrex tubes with cap

Preparation of the Samples

a) Micelle Preparation

L-a-lysolecithin is dispersed in 1.5 ml of distilled water at a concentration of 30 mg ml^{-1} and sonicated for 3 min at the maximum power. Part of this sample will be used as reference and the remaining to incorporate the GA.

b) Membrane Preparation

GA is added, as a powder, to 1 ml of lysolecithin micelles in water to the final concentration of about 11 mg ml^{-1}, so as to obtain a phospholipid to Gramicidin A molar ratio of 10:1.

The sample is then shaken for 3—4 min, sonicated for 6 min at the maximum power, and the appropriate volume of NaCl 100 mM is then added to a final salt concentration of 10 mM.

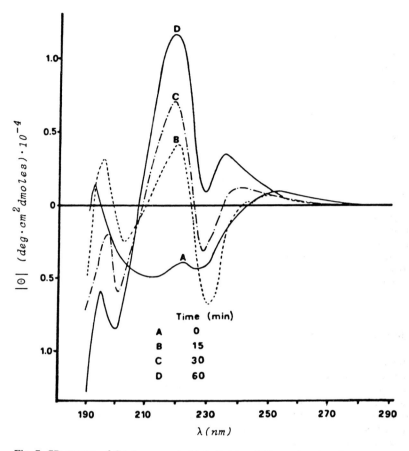

Fig. 7. CD spectra of GA incorporated into lipid at different incubation times

CD Measurements

The CD spectra are recorded with a Jasco 500A automatic recording spectro-polarimeter equipped with a DP-500N microprocessor unit and a thermosta-table cell holder.

0.2 mm optical pathlength cylindrical cells are used and the spectra are recorded after 0, 15, 30, 60 min of incubation at 70ºC.

In order to run the spectra samples (a) and (b) need to be diluted with 10 mM NaCl to a GA final concentration of about 1–2 mg ml^{-1}.

At the beginning of the incubation at high temperature there should not be any detectable spectrum as the GA is mostly in an aggregate state.

As a second step, a CD spectrum that can be attributed to a right handed helix should be recorded.

Finally, after about 1 h the CD spectrum should change to a pattern that has been associated to a single-stranded left-handed β-helix and that is considered to be stable conformation of the active channel (Fig. 7).

References

Hladky GB and Haydon DA (1972) Biochem Biophys Acta 274:294–312
Masotti L, Spisni A and Urry DW (1979) Cell Biophysics 2:241–247
Szabo G and Urry DW (1979) Science 203:55–57
Urry DW, Spisni A, Khaled MA and Long MM (1967) Int J Quantum Chem: Quantum Biol Symp 6:289–298
Urry DW (1971) Proc Natl Acad Sci USA 68:1907–1913
Urry DW, Long MM, Jacobs M and Harris RD (1975) Ann NY Acad Sci USA 274:203–210
Urry DW, Spisni A and Khaled MA (1979) Biochem Biophys Res Commun 88:940–949

Readings to implement the theoretical part:

Brahms J, Brahms S (1970) In: Fine structure of proteins and nucleic acids. Fasman GD, Timasheff SN (eds) Marcel Dekker, New York
Charney E (1979) The molecular basis of optical activity. John Wiley and Sons, New York
Cantor CR, Shimmel PR (1980) Biophysical Chemistry; Part II: Techniques for the study of biological structure and function. WH Freeman and Co, San Francisco
Bayley P (1980) Circular dichroism and optical rotation. In: An introduction to spectroscopy to biochemists. Brown SB (ed) Academic Press, London
Rousseau DL (1984) Optical techniques in biological research. Academic Press, New York
Campbell ID, Dweck RA (1984) Biological spectroscopy. The Benjamin Cummings Pub Co Inc, London

III. Protein Modification

Two Examples of Selective Fluorescent Labeling of SH-Groups with Eosin-5-Maleimide: The ADP/ATP Translocator and the Cytochrome c Oxidase Subunit III of Bovine Heart Mitochondria

MICHELE MÜLLER and ANGELO AZZI

I. Introduction

A. ADP/ATP Translocator

The ADP/ATP translocator is located in the inner membrane of mitochondria, where it catalyzes the vectorial exchange between cytosolic ADP and matrix ATP, a key process in the cellular energy supply of aerobic organisms (for review see Klingenberg 1980, Vignais 1976, Vignais et al. 1982). In beef heart mitochondria, the translocator is the most abundant integral protein, being about 10% of the total mitochondrial protein. The molecular weight of the monomeric translocator is about 32,500. The translocator has been extracted and purified as a dimeric carboxyatractylate (CAT) or bongkrekic acid complex by using Triton X-100. A complete amino acid sequence of the translocator revealed that the monomeric protein contains 4 cysteins out of the 297 amino acids. Sulfydryl groups are essential for the nucleotide translocation, since their hindrance or modification causes inhibition of the process. At least one sulfydryl group is masked when CAT is bound to the translocator. Eosin-5-maleimide (EMA, Fig. 1) reacts specifically with SH-groups and at low concentration does not permeate the inner membrane. In the absence of ADP the SH-groups of the ADP/ATP translocator do not react with N-ethylmaleimide (NEM), whereas all the other SH-groups located on the cytosolic side of the inner membrane react with NEM. The difference in reactivity between EMA an and NEM was used as an experimental approach to label selectively the ADP/ATP translocator in mitochondria, mitoplasts and submitochondrial particles (Müller et al. 1982, 1984). The aim of the experiment is to show the selectivity of the labeling and the effect of CAT and mersalyl on it.

B. Cytochrome C Oxidase Subunit III

Subunit III of cytochrome c oxidase is the second largest polypeptide of the whole complex (12 or 13 subunits) having a molecular weight of 29,918, as

Membrane Proteins, ed. by Azzi
©Springer-Verlag Berlin Heidelberg 1986

M. Müller and A. Azzi

Eosin−5−maleimide MW=742   **Fig. 1.** Molecular structure of eosin-5-maleimide

calculated from the DNA-inferred amino acid sequence (Anderson et al. 1982). This polypeptide has been suggested to be involved in the H^+-translocation function of the enzyme based on the following evidence. Inhibition of the H^+-pump by N, N'-dicyclohexylcarbodiimide (DCCD) has been found to be associated with a parallel labeling of subunit III (Casey et al. 1980), whose removal from the complex resulted also in the loss of the H^+-transfer activity (Pentillä 1983, Thelen et al. 1985). DCCD was shown to bind in the subunit III of the beef heart enzyme to the glutamyl residue 90 (Prochaska et al. 1981). This subunit contains also 2 cysteinyl residues (Cys 115 and 218), which are not evolutionarily conserved. Cys 115 of subunit III was shown to be located at the cytoplasmic face of the inner mitochondrial membrane and to be accessible to water-soluble SH-reagents such as iodoacetamide or dithionitrobenzoate (Malatesta and Capaldi 1982). On the other hand, it was not possible to label Cys 218 even after dissociation and partial denaturing of subunit III by sodium dodecylsulfate (SDS) (Verheul et al. 1982). EMA was used to selectively modify the cysteinyl residue 115 of subunit III, the only one which was shown to be reactive toward water-soluble SH-reagents (Müller and Azzi 1985).

C. Use and Properties of Eosin Derivatives

Eosin derivatives bound to proteins were mainly used for fluorescence studies ranging from the ps to the ms time windows (Cherry 1978, 1979, Johnson and Garland 1982, Müller et al. 1982, 1984, references cited in Haughland 1985). On the other hand eosin derivatives, in this particular case eosin-5-maleimide, were excellent tools (1) to analyze quantitatively the amount of SH-groups in proteins and (2) to determine their sidedness with respect to a membrane, since this reagent does not permeate the inner mitochondrial membrane up to 300 μM concentration (Houšték and Pedersen 1985), (3) to follow proteolytic degradation of the labeled protein by gel electrophoresis or chromatography, or (4) to monitor selective detachment of subunits from an intact complex (Müller and Azzi 1985), (5) the eosin-5-maleimide derivatization of cysteins survives also CNBr and 2-nitro-5-thiocyano-benzoic acid fragmentations allowing subsequent peptide analysis after gel electrophoresis (Müller unpublished results).

II. Experimental Procedure

A. Reagents and Buffers

1. Labeling Experiments

— Carboxyatractyloside (CAT) (Boehringer, Mannheim, FRG) 1 mg ml^{-1}
— N-ethylmaleimide, 1 mg ml^{-1}
— Mersalyl, 1 mg ml^{-1}
— Eosin-5-maleimide (EMA) (Molecular Probes, Junction City, OR, USA), dissolved in 100 mM sodium phosphate pH 7.2 and then diluted to 10 and 0.5 mg ml^{-1}
— Mercapto-1,3-propanediol
— Bovine serum albumin
— Bovine heart mitochondria, 10 mg ml^{-1}
— Isolated bovine heart cytochrome c oxidase prepared after Yu et al. 1975
— MSH-buffer: 220 mM mannitol, 70 mM sucrose, 10 mM 4-(2-hydroxyethyl)-1-piperazineethanesulfonic acid (Hepes)/KOH, pH 7.4
— 50 mM sodium phosphate, 1% sodium cholate, pH 7.2
— 20% sodium cholate, pH 7.2

2. Proteolysis of the ADP/ATP Translocator in Intact Mitochondria

— Trypsin, 1 mg ml^{-1} dissolved in 20 mM tris(hydroxymethyl)aminomethane (Tris)/Cl pH 8.5
— Trypsin-inhibitor, 1 mg ml^{-1} dissolved in 20 mM Tris/Cl pH 8.5

3. Isolation of the ADP/ATP Translocator

— 100 mM sodium sulfate, 10 mM 1,4-piperazinediethanesulfonic acid (Pipes), 0.05 mM EDTA, pH 7.2
— 20% (v/v) Triton X-100
— Hydroxylapatite Bio-Gel HTP (Bio-Rad)

4. Sodium Dodecylsulfate (SDS) Polyacrylamide Gel Electrophoresis

Final acrylamide concentration	13.5%	5%
30% (w/v) acrylamide, 0.8% (w/v) bisacrylamide	9 ml	1.3 ml
1.5 M Tris/Cl, 8 mM EDTA, 0.4% (w/v) SDS, pH 8.8	5 ml	—
0.5 M Tris/Cl, 8 mM EDTA, 0.4% (w/v) SDS, pH 6.8	—	2.0 ml
3% (w/v) Polyacrylamide (EGA-Chemie, Steinheim, FRG)	3.3 ml	1.3 ml
Water	2.6 ml	3.3 ml
N,N,N',N'-tetramethylethylenediamine	10 μl	5 μl
10% (w/v) ammonium persulfate	100 μl	160 μl

The above quantities are given for a 130 x 100 x 1.5 mm slab gel.

- Electrode buffer: 50 mM Tris, 380 mM glycine, 0.1% (w/v) SDS, 1.8 mM EDTA
- Sample buffer: 10 mM sodium phosphate pH 7, 10% (v/v) glycerol, 2.5% (w/v) SDS, 10 mM mercapto-1,3-propanediol, (0.5% bromphenolblue)
- Fixing solution: 25% (v/v) ethanol, 14% (v/v) formaldehyde
- Staining solution: 0.25% (w/v) Coomassie blue, 50% (v/v) methanol, 7.5% (v/v) acetic acid
- Destaining solution: 20% (v/v) methanol, 7.5% (v/v) acetic acid
- Vaseline or 0.5% (w/v) Agar-agar solution

B. Equipment

- Ultracentrifuge, rotor with small (1–5 ml) tubes
- Eppendorf tubes, supports and centrifuge
- Ice bucket with lid
- Vortex mixer
- Stop watch
- Complete vertical slab gel electrophoresis set with power supply
- Spectrophotometer
- UV light box
- Polaroid camera and film (Type 665) with UV and orange or yellow filters
- 100 μl microsyringe

C. Labeling of the Translocator

All experiments are performed on ice and the EMA solution should not be exposed to light until the reaction is arrested with mercapto-1,3-propanediol. The basic experiment is performed as follows: to 1 ml of bovine heart mitochondria (10 mg ml^{-1}) 15 μl of EMA (10 mg ml^{-1}) are added and immediately mixed. After 20 min of incubation in the dark, 10 μl of mercapto-1,3-propanediol are added to the mitochondria and further incubated for 15 min in the dark (Fig. 2, lane 1). Free EMA is removed by several centrifugations (each of 1 min) using an Eppendorf centrifuge until the supernatant is clear. Addition of bovine serum albumin to the MSH-buffer helps to accelerate the removal of free EMA. To obtain the selective labeling of the translocator the mitochondria must be pretreated with 100 μl NEM (1 mg ml^{-1}) for 5 min before EMA labeling (Fig. 2, lane 2). To test the inhibitory effect of mersalyl, also 5 min preincubation of the mitochondria with this mercurial at the same concentration as NEM is needed. CAT (4 μg mg^{-1} mitochondrial protein; 20 min preincubation with freshly isolated mitochondria) is used to demonstrate the specificity of the described labeling procedure for the ADP/ATP translocator. Also in this case CAT-labeled mitochondria are treated with NEM as described before, prior to their labeling with EMA (Fig. 2, lane 3).

D. Trypsinisation of EMA-Labeled Translocator in Mitochondria

To about 0.5 mg of mitochondria selectively labeled with EMA in 50 μl of MSH-buffer (pH 8), an aliquot of 6 μl of freshly prepared trypsin solution (1 mg ml^{-1}) was added and incubated at 20°C. After 30 s, 1, 2, and 3 min, 10 μl aliquots were rapidly transferred into Eppendorf tubes containing each 20 μl of hot sample buffer (90°C) and 5 μl of trypsin-inhibitor (1 mg ml^{-1}). The trypsinized samples were then directly analyzed by SDS polyacrylamide gel electrophoresis (Fig. 3).

E. Isolation of the EMA-Labeled Translocator

The EMA-labeled mitochondria (10 mg) are centrifuged and the pellet resuspended in 1 ml of 100 mM sodium sulfate, 10 mM Pipes, 0.05 mM EDTA, pH 7.2 and Triton X-100 to a final concentration of 6% (v/v). After 30 min at 4°C the solution is centrifuged at 140,000 g for 30 min. The supernatant is then absorbed on a hydroxylapatite column (Pasteur pipet, containing about 600 mg of dry material) and eluted with detergent-free buffer until the pink fraction (EMA-labeled ADP/ATP translocator) is eluted. 50 μl of

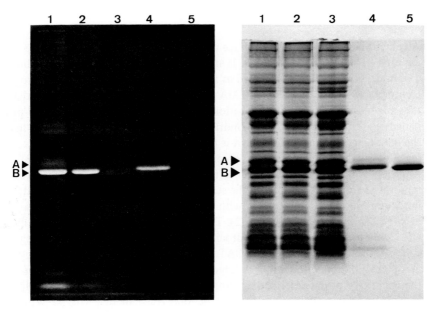

Fig. 2. SDS-polyacrylamide gel electrophoresis of inner mitochondrial membranes (IMM) and purified ADP/ATP translocator after labeling with EMA. **A** Fluorograph of the gel before staining with Coomassie blue. Labeling of IMM was performed as described under "Experimental procedures". Lane *1* is the result of the labeling without any pretreatment. Lane *2* shows the effect of the NEM pretreatment resulting in the selecitve labeling of the ADP/ATP translocator. Lane *3* demonstrates the very selective inhibitory effect of CAT on the EMA labeling. Lane *4* represents the isolated ADP/ATP translocator isolated from EMA-labeled mitochondria (Band B). **B** Coomassie blue stain of the same gel. Purified CAT-bound ADP/ATP translocator was co-electrophoresed (lane *5*) with the isolated EMA-labeled protein showing good agreement between the two migrations. Band *A* shows the position of the phosphate transport protein contaminating the preparation of lane *4* which, however, was not fluorescent

this fraction are separated on SDS polyacylamide gel, which fluorography (Fig. 2A, lane 4) reveals the presence of EMA fluorescence associated with the translocator. Contaminants of this crude preparation visible in the protein stain picture (Fig. 2B, lane 4) are not fluorescent.

F. Labeling of Cytochrome C Oxidase Subunit III

Cytochrome c oxidase is labeled with EMA as follows: 1 nmol mersalyl is added to 1 nmol cytochrome c oxidase in a buffer composed of 50 mM sodium phosphate (pH 7.2), 1% (w/v) sodium cholate, and incubated for 30 min on

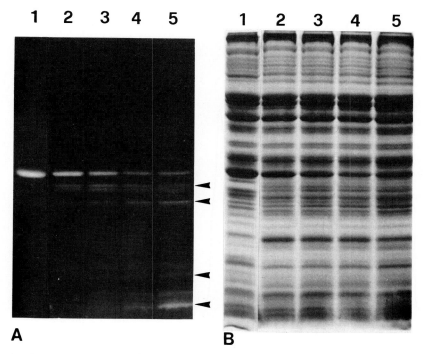

Fig. 3. SDS-polyacrylamide gel electrophoresis of EMA-labeled ADP/ATP translocator trypsinized in intact mitochondria. **A** Fluorograph of the gel before staining with Coomassie blue. Labeling and trypsinization were performed as described under "Experimental procedures". Lane *1* selectively labeled ADP/ATP translocator in mitochondria not treated with trypsin. Lanes *2, 3, 4, 5* Trypsinization of the EMA-labeled translocator was blocked after 30 s, 1, 2, and 3 min respectively. The trypsinization proves the formation of fluorescently labele degradation peptides (*arrows*) allowing their identification between the multitude of protein visibles on the Coomassie blue stain of the same gel (**B**)

ice; then 1 nmol of EMA per nmol oxidase is added to the pretreated enzyme and incubated for 2 h on ice in the dark. To label all free SH-groups of cytochrome c oxidase, the enzyme (1 nmol) is dissolved in 50 μl of sample buffer for 30 min, at room temperature prior to its labeling with 30 nmol of EMA. The reaction is stopped by adding 1 μl of mercapto-1,3-propanediol (Fig. 4).

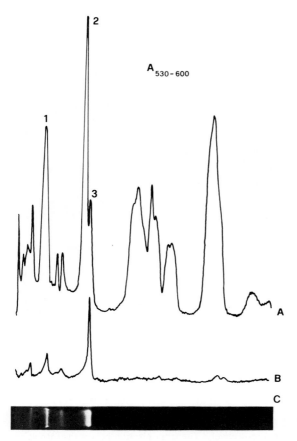

Fig. 4. SDS polyacrylamide gel electrophoresis of cytochrome c oxidase labeled with EMA. **A,B** Densitometric traces of the slab gel measured immediately after electrophoresis without any fixation. EMA absorbance was recorded at the wavelength pair 530–600 nm with a special attachment for the Amino DW-2a spectrophotometer. Each lane contained about 1 nmol of enzyme. *Numbers* indicate the subunit sequence. In **A** cytochrome oxidase was dissolved in sample buffer prior to its labeling with EMA. In **B** the labeling was performed in 1% cholate as described under "Experimental procedures". **C** Fluorography of **B** taken before the densitometric analysis

G. Quantitative Determination of Bound EMA

1. ADP/ATP Translocator

100 μl (1 mg) of each sample are diluted to 1 ml with MSH buffer in the presence of 5% sodium cholate. Spectra are recorded from 500 to 650 nm and the absorption value at 600 nm is subtracted from that at 530 nm. The ab-

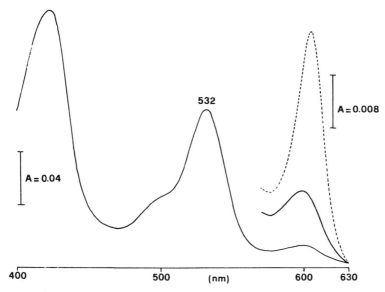

Fig. 5. Absorption spectra of EMA-labeled cytochrome oxidase. *Solid line* air-oxidized enzyme; *dotted line* dithionite-reduced enzyme. The central peak is the absorption maximum of EMA (532 nm). The EMA/enzyme molar ratio was calculated to be about 1

sorption values have to be corrected for the mitochondrial protein background absorption by using a suspension of the same protein content. The amount of EMA bound to the membrane is calculate using the absorption coefficient $\Delta\epsilon(530-600) = 83$ mM^{-1} x cm^{-1}. The EMA/dimeric translocator molar ratio is calculated knowing that per mg mitochondrial protein there are 1.3– 2.0 nmol of dimeric ADP/ATP translocator.

2. Cytochrome c Oxidase

The heme a content is calculated from spectra (dithionite-reduced minus air-oxidized) using the extinction coefficient $\Delta\epsilon(604-630) = 13.5$ mM^{-1} x cm^{-1}. EMA content is calculated as described above. Usually 1 nmol of labeled oxidase is diluted to 1 ml 50 mM sodium phosphate, 1% sodium cholate, pH 7.2 (Fig. 5).

H. SDS Polyacrylamide Gel Electrophoresis

A 13.5% polyacrylamide slab gel is prepared and assembled on a gel electrophoresis vessel filled with electrode buffer. From each mitochondrial sample

10 μl (about 100 μg protein) are mixed with 30 μl water and 20 μl of sample buffer (without bromphenol blue) in an Eppendorf tube. From each cytochrome c oxidase sample 1 nmol of enzyme is mixed with 1 vol of sample buffer. To denaturate the proteins the samples are left for 30 min at room temperature and then applied to the gel. The electrophoresis should be run in the dark at 30 mA constant current. After electrophoresis the gel without glass plates is illuminated by a UV light source visibilizing the EMA fluorescence. The photograph is taken through UV and orange or yellow filters. The gel is subsequently fixed for 1 h, stained for 1 h and destained overnight by conventional procedure.

References

Anderson S, DeBrujn MHL, Carlson AR, Eperon IC, Sanger F, Young IG (1982) J Mol Biol 156:683–717
Casey RP, Thelen M, Azzi A (1980) J Biol Chem 255:3994–4000
Cherry RJ (1978) Methods Enzymol 54:47–61
Cherry RJ (1979) Biochim Biophys Acta 559:289–327
Haughland RP (1985) Handbook of fluorescent probes and research chemicals. Molecular Probes, Junction City, OR 97448, USA
Houštěk J, Pedersen PL (1985) J Biol Chem 260:6288–6295
Johnson P, Garland PB (1982) Biochem J 203:313–321
Klingenberg M (1980) In: Lee CP, Schatz G, Dallner G (eds) Mitochondria and microsomes. Addison-Wesley, Reading, Mass, pp 293–316
Malatesta F, Capaldi RA (1982) Biochim Biophys Res Commun 109:1180–1185
Müller M, Azzi A (1985) FEBS Lett 184:110–114
Müller M, Krebs JJR, Cherry RJ, Kawato S (1982) J Biol Chem 257:1117–1120
Müller M, Krebs JJR, Cherry RJ, Kawato S (1984) J Biol Chem 259:3037–3043
Pentillä T (1983) Eur J Biochem 133:355–361
Prochaska LJ, Bisson R, Capaldi RA, Steffens GMC, Buse G (1981) Biochim Biophys Acta 637:360–373
Thelen M, O'Shea PS, Petrone G, Azzi A (1985) J Biol Chem 260:3626–3631
Verheul FEAM, Draijer JW, Muijers AO, Van Gelder BF (1982) Biochim Biophys Acta 681:118–129
Vignais PV (1976) Biochim Biophys Acta 456:1–38
Vignais PV, Block MR, Boulay F, Brandolin G, Lauquin GMJ (1982) In: Martonosi AN (ed) Membrane and transport, vol I, Plenum Press, New York, pp 405–413
Yu C, Yu L, King TE (1975) J Biol Chem 250:1383–1392

Hydrophobic Photolabeling with ^{125}I-TID of Red Blood Cell Membranes

C. MONTECUCCO

I. Introduction

Reagents such as 3-trifluoromethyl, 1—3(m-iodophenyldiazirine) (TID) have been introduced with the aim of identifying integral protein in a membrane and to study the hydrophobic sector (Bayley 1983). To reach this goal the reagents have been made:

1. Hydrophobic, in order to make them readily partition in the hydrophobic domain of the membrane; this property will determine their localization (however, by the same property they will also be able to occupy hydrophobic pockets present on the hydrophilic domain of an integral protein);

2. Photoactivatable, in such a way that they can be activated in a very short time only when all the biological manipulations are ended. This most important property of photolabels allows one to start the labeling reaction only when needed; moreover in this way intermediates can be generated with such a reactivity that they will be able to form covalent derivatives even with the aliphatic side chains frequently found in the lipid-exposed surface of integral proteins;

3. Radioactive, in order to be able to trace the labeling at the level of the protein or the subunit (mainly by SDS-gel electrophoresis as in the present experiment) and even further at the level of the modified residues (by sequencing). In this sense ^{125}I is one of the best isotopes to be used because of its high specific radioactivity, which lowers the amount of probe needed (and parallely the perturbation of the system), and because of its easier detectability in autoradiography.

The results obtained must be interpreted with care in order to exclude the possibility that hydrophilic or peripheral proteins are labeled, since they may contain hydrophobic pockets, which can be occupied by the probe. Moreover, it must be considered that, although very reactive, the photogenerated intermediates are not unspecific and hence a protein may not be labeled be-

Membrane Proteins, ed. by Azzi
© Springer-Verlag Berlin Heidelberg 1986

cause of the nature of the residues forming its hydrophobic sector. These and other points concerning the potential and the possible drawbacks and artifacts of the hydrophobic photolabeling method and the kind of information which can be obtained with the reagent described here and with more sophisticated ones (such as photoreactive phospholipid analogs) are discussed in a recent review (Bisson and Montecucco 1985).

The best of the small lipophilic photoreactive probes is [^{125}I]-TID because of its high reactivity, specific radioactivity, and commercial availability from Amersham International (U.K.) cat. IM. 148 (Brunner and Semenza 1981). For the above reasons this reagent has been chosen for the present experiment on labeling the red blood cell membrane.

Red blood cells will be incubated with TID, photolyzed with long-wave, nonprotein-damaging, ultraviolet radiations, recovered in a small pellet and, after dissolution in SDS, electrophoresed. The radioactivity associated with the protein and lipid bands will be determined by autoradiography of the dried gel. The pattern of labeling will provide us with some information on the localization of the different proteins with respect to the membrane.

II. Experimental Procedure

A. Materials

— 150 mM NaCl, 5 mM NaPi, pH 7.3 (buffer 1)
— 10% Sucrose, 5 mM NaPi, pH 7.3 (buffer 2)
— BSA (bovine serum albumin) 1.5% in buffer 1
— Human erythrocyte membranes (resealed ghosts), 0.5 mg ml^{-1} in buffer 2
— 3-Trifluoromethyl 1-3-(m-[^{125}I] iodophenyl) diazirine (^{125}I-TID), ethanol solution as purchased from Amersham
— Ultracentrifuge with swing out rotor and plastic tubes
— Glass plates with spacers, clamps, gel electrophoresis apparatus and power supply
— Acrylamide 30%, bis-acrylamide 0.8%
— Tris-Cl 2 M, pH 8.8
— Tris-Cl 0.5 M, pH 6.8
— SDS Na-dodecylsulfate 20%
— TEMED (N,N,N', tetramethylethylenediamine)
— Ammonium persulfate 10%
— Sample buffer: 4% SDS, 10 mM Tris-Acetate, 0.1 mM EDTA, pH 8.2; β-mercaptoethanol is added to final 3% just before use together with 5% of a concentrated solution of Bromo-Phenol Blue in 70% glycerol

— Running buffer: for 1000 ml 6 g Tris base, 2.88 g glycine, 1 g SDS, pH 9.2
— microsyringes, pipets, vials, tubes UV lamp, γ-counter, gel dryer, films, film casset and developing and fixing solution.

B. Methods

— 260 μl of RBC membrane suspension are transferred with a pipet in an ir-radiation glass vial in ice, flushed with nitrogen and the vial is stoppered
— In a dark fume cabinet 1–2 μl of the TID solution are taken with a micro-syringe and added to the membranes
— After a 20-min incubation with occasional hand stirring, the sample is placed on a ice-filled small Becker and the Becker on a UV-lamp turned upside down. The lamp is switched on for 30 min
— 130 μl of the BSA solution are then added, the suspension stirred and in-cubated in ice for 20 min
— The sample is layered on a 5 ml Beckman transparent plastic tube con-taining 4.2 ml of buffer 2. After carefully balancing the tubes, the centri-fuge is started and centrifugation performed at 100,000 g for 60 min
— During centrifugation, a SDS-polyacrylamide gel according to a slightly modified version of the Laemmli procedure is prepared
— Assemble the glass plates with spacers and clamps
— In two different Beckers add the following solutions one after the other given, under stirring

	Running gel	Step gel
Acrylamide 30% Bis-acrylamide 0.8%	8.3 ml	1.33 ml
Tris-Cl 2 M, pH 8.8	5 ml	–
Tris-Cl 0.5 ml, pH 6.8	–	1.5 ml
SDS 20%	62.5 μl	50 μl
TEMED	30 μl	30 μl
Sucrose	–	1.5 gr
— Add water to a final volume of	25 ml	10 ml

— Place the comb, add 60 μl of 10% ammonium persulfate to the running gel solution and after 2 min pour it in between the two glass plates till 1 cm from the comb. Add a few dorps of water-saturated isobutanol. After polymerization is accomplished, remove isobutanol by absorption with filter paper, wash with water and remove it

— Add 30 μl ammonium persulfate to the step gel solution and after 2 min of stirring transfer the latter with a pipet to both sides of the plates (very slowly and with great care) on top of the running gel. The gel should polymerize in less than 1 h
— Pour out the supernatant of the centrifuged sample and add 30 μl of sample buffer, vortex, and seal the top with aluminum paper and parafilm. Boil for 3 min. Cool in a water bath, add 10 μl of running buffer and vortex thoroughly
— Mount the glass plates on the gel holder, fill the running buffer reservoirs and load the sample with a microsyringe on the gel. Set the power supply to 8 mA constant current until the entire sample has entered the gel and then rise the current to 25 mA constant current until the tracking dye Bromo Phenol Blue has run out of the gel.
— Place the gel for 10 min in a 227 ml methanol, 227 water, 46 ml acetic gel fixing solution.
— Transfer the gel into a 0.25% Coomassie Brilliant Blue solution in 30% methanol, 7.5% acetic acid for 1 h. Place the gel in a destaining solution until removal of the Blue background.
— The result obtained at this point is very similar to that described by Steck (1974) in terms of protein pattern. The spectrins, band III, and some glycophorins are clearly identified. It is evident that the present gel is unable to resolve all the components of the erythrocyte membrane. A much better resolution can be obtained with a gradient gel, where all the bands listed by Bennett (1985) are clearly separated
— Dry the gel in a gel-drying apparatus at 80ºC under the vacuum of a water pump for 2 h
— Autoradiography and fluorography of radioactive gels is simply and clearly described in Amersham review no. 23 by R.A. Laskey. Place the gel in a film casset provided with a screen for ionizing radiation together with a Kodak X-OMAT AR film in between the gel and the screen. Close, wrap in aluminum foil and store at -78ºC for the time needed (from half a day to a few days depending on the amount of radioactivity used).
— Let the casset warm up to room temperature and develop the film in the dark in a developing solution for 60–90 s and then transfer the film to a fixing solution for 15 min (red light now on). Wash thoroughly with water, then with distilled water and let it dry in a dust-free place.
— By comparing the coomassie Blue staining pattern with the autoradiographic one, it can be seen that, while the peripheral protein spectrin is not labeled, most of the radioactivity is associated with band III and with the glycophorins, as expected from their integral nature.

References

Bayley H (1983) Photogenerated reagents in biochemistry and molecular biology. Elsevier Biomedical Press, Amsterdam

Bennett V (1985) Annu Rev Biochem 54:273–304

Bisson R, Montecucco C (1985) In: Watts A, De Pont JJHHM (eds) Progress in lipid-progress interaction, Elsevier Biomedical Press, Amsterdam, pp 259–287

Brunner J, Semenza G (1981) Biochemistry 20:7174–7182

Steck TL (1974) J Cell Biol 62:1–19

Use of Fluorescent Probes of the Adenine Nucleotide Carrier for Binding Studies and Analysis of Conformational Changes

GÉRARD BRANDOLIN, MARC R. BLOCK, FRANÇOIS BOULAY and
PIERRE V. VIGNAIS

I. Introduction

The adenine nucleotide carrier (AdN carrier) is a membrane-bound protein
(minimum $M_r \simeq 32,000$) which is responsible for the vectorial exchange between mitochondrial ATP and cytosolic ADP through the inner mitochondrial
membrane. During the course of oxidative phosphorylation, the ATP synthesized in mitochondria is exported against ADP generated in cytosol by energy-
consuming reactions; this asymmetric exchange is governed by the membrane
potential generated by mitochondrial respiration. Two specific inhibitors of
the AdN carrier, (carboxy) atractyloside (CATR) and bongkrekic acid (BA),
have been used extensively to study the mechanism of ADP/ATP transport.
The fact that CATR binds to the carrier from the outside of the mitochondrial
membrane and BA from the inside points to the asymmetric orientation of the
carrier in the membrane (for review, see Vignais et al. 1985).
 The AdN carrier molecule is able to adopt two different conformations
referred to as CATR and BA conformations, which react with CATR and BA
to form carrier-inhibitor complexes. For the membrane-bound AdN carrier,
and the isolated AdN carrier in detergent, it has been shown that the two conformations are in equilibrium and that transportable nucleotides are able to
trigger the transition between the two conformations (Block et al. 1983, Brandolin et al. 1981, 1985). This has led to the suggestion that the transition between the CATR and BA conformations has a physiological significance in
terms of transport. The denomination cytosolic state or c state and matrix
state or m state (Klingenberg 1981) is based on the postulate that a single
nucleotide site per carrier dimer moves alternatively between the cytosol and
the matrix space. As the c state and the m state are believed to correspond to
two conformations of the AdN carrier during transport, and since CATR binds
to the AdN carrier by the cytosolic face and BA by the matrix face, it is clear
that the c and m states are equivalent to the CATR and BA conformations respectively. However, because of controversies which still exist concerning the
nucleotide binding site(s) and recent evidence indicating the presence of several interacting nucleotide binding sites in the AdN carrier molecule (Dupont

Membrane Proteins, ed. by Azzi
©Springer-Verlag Berlin Heidelberg 1986

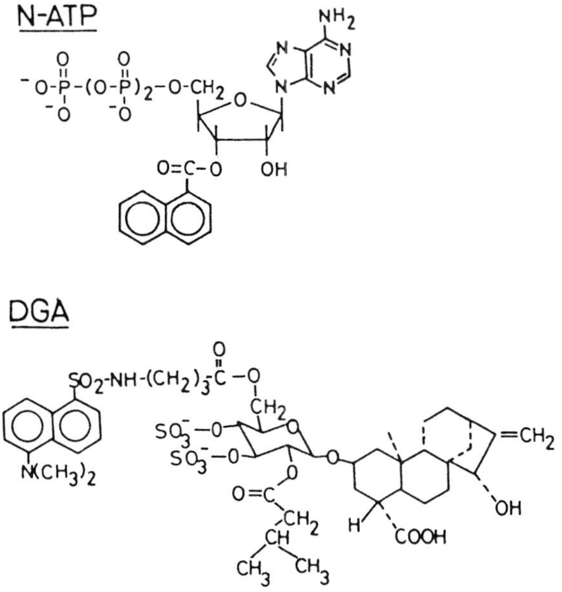

N-ATP

DGA

Fig. 1. Structure of 3'-0-naphthoyl-ATP (N-ATP) and of dansyl-γ-aminobutyryl atractyloside (DGA)

et al. 1982, Brandolin et al. 1982, Block and Vignais 1984), we prefer to keep the operational denominations of CATR and BA conformations.

In the experiments to be described here, intrinsic and extrinsic fluorescent probes are used to study the ATP- or ADP-induced transition between the CATR and the BA conformations. The intrinsic fluorescence of the AdN carrier is essentially due to tryptophan (5 tryptophanyl residues for a minimum M_r of 32,000) (Aquila et al. 1982). Intrinsic fluorescence assays are carried out with the beef heart AdN carrier protein isolated in detergent to avoid the contribution of tryptophanyl groups from other mitochondrial proteins (Brandolin et al. 1981, 1985). Aminoxide-type detergents are used for this study.

The extrinsic fluorescence experiments involve the use of the nucleotide analogs, 3'-0-naphthoyl ADP and 3'-0-naphthoyl ATP (Block et al. 1982, Dupont et al. 1982, Block et al. 1983) and the inhibitor analog: the 6'-0-dansyl-aminobutyryl atractyloside (Boulay et al. 1983).

These specific fluorescent ligands (Fig. 1) are able to probe the binding sites of the AdN carrier and they behave as reporters of the conformational states assumed by the AdN carrier.

A. Reagents and Buffers

- Beef heart mitochondria, 50 mg ml^{-1} in 0.25 M sucrose, 10 mM Tris-HCl final pH 7.4 (Smith 1967)
- 3-laurylamido-NN'-dimethyl propylaminoxyde (LAPAO) synthesized as previously described (Brandolin et al. 1980). If not available LAPAO can be substituted by aminoxide WS 35 (from Theo Goldschmidt A.G., Essen, FRG), which is a mixture of LAPAO (55%) and longer chain aminoxides.
- 1 mM and 10 mM ADP pH 7.4
- 1 mM and 10 mM ATP pH 7.4
- Atractyloside (ATR) (Sigma) (1 mM solution in water)
- Carboxyatractyloside (Boehringer), 1 mM solution in water
- Bongkrekic acid (BA) (1 mM solution in water, after neutralization by ammonia). BA was obtained by the method of Limjback et al. (1970), as modified by Lauquin et al. (1976).
- 6'-0-dansyl-4-aminobutyryl-atractyloside (DGA). The synthesis of DGA is described hereafter.
- Hydroxyapatite (Biogel HTP–BioRad)
- 1-Naphthoic acid (Fluka)
- 1-1'-carbonyl-bis-imidazole (Merck)
- Naphthoyl-ADP and -ATP (N-ADP and N-ATP). The synthesis of N-ADP and N-ATP is described hereafter
- Dansyl chloride (Pierce)
- HTP buffer (0.1 M Na_2SO_4, 10 mM tricine-KOH, 0.1 mM EDTA, 0.5% W/V LAPAO, final pH 7.2).
- AcA buffer (50 mM MOPS, 0.1 mM EDTA, 0.5% LAPAO, final pH 7.0)
- 136 mM glycerol
- Buffer for N-ADP assay (120 mM KCl, 1 mM EDTA, 10 mM MES, final pH 6.5)
- Concentrated lysis medium kept at room temperature (0.5 M Na_2SO_4, 50 mM tricine-KOH, 0.5 mM EDTA, 5% LAPAO, final pH 7.4).
- Buffer for DGA assay (250 mM sucrose, 1 mM EDTA, 10 mM HEPES, final pH 7.0)
- 1 and 10 mM UDP-UTP pH 7.4
- 1 and 10 mM CDP-CTP pH 7.4
- 1 and 10 mM GDP-GTP pH 7.4
- 1 and 10 mM AMP pH 7.4

B. Equipment

- High sensitivity fluorimeter (Biologic Co, ZIRST, 38240 Meylan, France) equipped with a stirring device and a thermostated cuvette holder

— Optical filters K1 (410 nm) K50 (520 nm) from Baltzers and 0—52 or 7—54 from Corning.
— 3 ml quartz fluorescence cuvette, 1 cm path
— Hamilton automatic syringes (20 μl)
— HPLC apparatus with a preparative μBondapak C18 column (Waters-Millipore)
— 25,000 g centrifuge
— Vortex mixer

II. Preparation of Purified ADP/ATP Carrier Protein

Suspend 25 g dry hydroxyapatite in 6 volumes of an ice-cold medium consisting of 0.1 M Na_2SO_4, 0.1 mM EDTA and 0.5% LAPAO w/v, 10 mM Tricine-KOH buffer, final pH 7.2. After decantation, the supernatant and fines are carefully eliminated. Repeat this step once with one volume of medium. Resuspend the gel in one volume of medium and pour the suspension in a 2.5 cm x 30 cm column to a height of 10 cm (50 ml decanted HTP). To an 8-ml fraction of a diluted suspension of beef heart mitochondria (10—12 mg protein ml^{-1}) add 2 ml of a detergent medium which consists of 0.5 M Na_2SO_4, 50 mM tricine-KOH, 0.5 mM EDTA and 5% LAPAO (w/v), final pH 7.4. After a 5-min incubation in ice, the lyzed mitochondria are centrifuged at 25000 g for 30 min. The supernatant is carefully withdrawn and layered on the hydroxyapatite, and monitored by UV absorption.

Collect the first protein peak which is eluted just after the void volume. This fraction contains about 0.5 mg protein ml^{-1}. It constitutes a stock preparation of solubilized AdN carrier protein which can be stored in liquid nitrogen in 5 to 10 ml aliquots.

Prior to each experiment, the AdN carrier protein is further purified by chromatography on an AcA 202 gel column. This step also allows the removal of a significant amount of Ca^{2+}. Layer a 10-ml fraction of the hydroxyapatite carrier preparation on a 40-ml AcA 202 column previously equilibrated in a medium consisting of 50 mM MOPS, 0.1 mM EDTA, 0.5% LAPAO (w/v) final pH 7.00. The protein fraction is eluted just after the void volume of the column and is kept at 0°C.

III. Synthesis of Naphthoyl-ADP and Naphthoyl-ATP (N-ADP and N-ATP)

The method used is based on the esterification of the 3' alcohol group of the ribose moiety of ADP or ATP by a carboxylic acid as described by Gottikh et al. (1970) and adapted by Guillory (1979). 1-Naphthoic acid used in the present preparation is first activated by the carbonyl reagent: 1-1'-carbonyl bis imidazole.

Add 1-naphthoic acid (0.1 mmol, 17 mg) and 1-1'-carbonyl bis-imidazole (0.3 mmol, 50 mg) to 0.1 ml of anhydrous dimethylformamide (freshly distilled and kept on calcium hydride). The mixture is stirred for 30 min in darkness prior to the addition of ADP or ATP (25 μmol dissolved in 0.5 ml of water). The reaction is allowed to proceed for 2 to 3 h at room temperature. The solvents are evaporated under vacuum in a rotatory evaporator at 40°C. Extract the gummy residue three times with 5 ml of dry acetone to remove the excess of naphthoic acid and imidazole. Dissolve the final residue in 1 ml of a 1/1 (v/v) mixture of ethanol and water and apply it to a 23 x 57 cm sheet of Whatman 3 paper for a descending chromatography.

The chromatogram is developed for 15 h at room temperature with the following system: n-butanol, acetic acid, water (5/2/3 v/v/v). N-ADP and N-ATP are localized under UV light by their deep ultraviolet fluorescence; their Rf values are 0.62 and 0.51 respectively. Cut out the paper strips corresponding to the derivatives and elute N-ADP or N-ATP with water by descending chromatography. The completion of elution is monitored under UV light.

The molecular extinction coefficients of both N-ADP and N-ATP at pH 7.00 are 15400 M^{-1} cm^{-1} at 260 nm and 6200 M^{-1} cm^{-1} at 300 nm. The two derivatives are characterized by a fluorescence excitation peak at 310 nm and a fluorescence emission peak at 395 nm.

IV. Synthesis of 6'-0-Dansyl-4-Aminobutyryl-Atractyloside (DGA)

The principle of this synthesis is the esterification of a primary alcohol group situated at the 6' position of the glucose moiety of the atractyloside molecule by dansyl-4-aminobutyric acid (Boulay et al. 1983). If not available, the dansyl-4-aminobutyric acid can be synthesized as follows. A solution of dansyl chloride (600 μmol) in 2 ml of anhydrous acetone is added dropwise with stirring to a solution of 4-aminobutyric acid (300 μmol) dissolved in a mixture of pyridine and water (1/1 v/v).

The pH value is maintained at a value of 9 by addition of triethylamine. The solution is stirred for 12 h at room temperature. After removing the solvents under reduced pressure at 30°C, the residual solid is dissolved in 5 ml of water and the pH is brought to a value of 2 by addition of 2N HCl. Dansyl aminobutyric acid is extracted three times by 10 ml of ethyl acetate. The pooled extracts are washed with 20 ml of saturated NaCl solution and then the solvent is evaporated under reduced pressure at 30°C. The residue is solubilized in 0.01M NaOH and washed by diethylether. Finally, the aqueous phase which contains the pure dansyl-4-aminobutyric acid is neutralized. After drying, 10 µmol of dansyl-4-aminobutyric acid are dissolved in anhydrous pyridine (1 ml) and allowed to react with a stoichiometric amount of thionyl chloride for 3 min to form the dansyl-4-aminobutyryl chloride. ATR (10 µmol) dissolved in 2 ml of anhydrous pyridine is then added. The reaction of condensation is allowed to proceed overnight at 4°C. Pyridine is removed in a rotatory evaporator at room temperature and the unreacted dansyl-4-aminobutyric acid is extracted from the residue by repeated washings with anhydrous acetone. The final gummy product is dissolved in 2 ml of methanol-water mixture (1/1, v/v), filtered through a Millex FG filter and applied to a µBondapak C18 column (300 x 7.8 mm, 10 µm) for HPLC. The elution is performed with a linear gradient of methanol in water from 0 to 100%, always in the presence of 1% (v/v) acetic acid and 1% (v/v) 1M ammonium acetate. Pure dansyl-4-aminobutyryl atractyloside (DGA) is eluted at a concentration of methanol of 90%.

DGA is identified after chromatography on a Whatman K6 plate in chloroform/methanol/acetic acid/water (60/20/0.5/0.5, v/v/v/v) as a fluorescent spot with an Rf value of 0.45.

V. Use of N-ADP and N-ATP as Fluorescent Probes of the Conformational State of the Membrane-Bound AdN Carrier

A. Principle

N-ADP and N-ATP bind to mitochondria; however, they are not transported. They competitively inhibit ADP/ATP transport. Through the use of radiolabeled N-ADP or N-ATP, it has been shown that bound N-ADP or N-ATP are released by a saturating concentration of CATR (Block et al. 1982, 1983). The binding parameters for these analogs are as follows; number of sites: 1.4–1.6 nmol mg^{-1}; Kd value close to 3 µM. In fluorescence assays, the release of N-ADP or N-ATP induced by addition of CATR is accompanied by a fluorescence increase. The fluorescence emission of N-ADP or N-ATP is measured at 410 nm through a K1 (Baltzers) filter, with an excitation light centered at 310 nm.

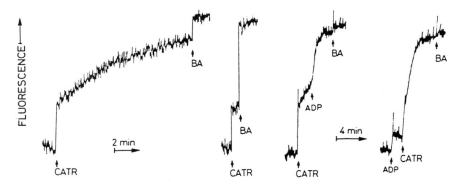

Fig. 2. Fluorescence increase (corresponding to the release of bound N-ADP) induced by CATR, BA and ADP (for details see text)

B. Assay

Introduce an aliquot of the suspension of beef heart mitochondria corresponding to 2 mg protein in the fluorescence cuvette which contains 2.5 ml of a saline
medium made of 120 mM KCl, 1 mM EDTA, 10 mM MES, pH 6.5 with 5 μM
N-ADP (Fig. 2).

The temperature is maintained at 10°C. When fluorescence has attained a
steady level, indicating complete equilibrium between bound and free N-ADP
(which requires 2 to 3 min), 10 μl of 1 mM CATR is added. A rapid increase
in fluorescence is observed, which lasts for less than 1 s. This rapid increase is
followed by a slower one which reaches a steady level in about 20 min. Another addition of CATR has no effect on the fluorescence signal, which points to
a specific and saturating effect.

Repeat the experiment starting by the addition of CATR. Add 10 μl of
1 mM BA after the rapid increase in fluorescence is achieved and at the onset
of the slow phase. Addition of BA induces a rapid increase of fluorescence
up to a plateau. Control the absence of effect of a second addition of BA.

Repeat the same experiment starting with CATR and replacing BA by
ADP (10 μl of 1 mM ADP). The effect produced by the addition of ADP in
the presence of CATR has to be compared with the effect of ADP on the N-
ADP release in the absence of CATR. The effect of ADP alone is very weak
compared to that obtained in the preceding experiment and is probably attri-
butable to a slight competition of ADP and N-ADP for binding to the AdN
carrier. The subsequent addition of CATR produces maximal effect in fluorescence increase.

Repeat the experiments concerning the N-ADP release by adding first BA
and then CATR or ADP (same amounts as described above).

C. Interpretation

The large effect of ADP observed in the presence of CATR and BA cannot be due to a competitive effect, since ADP alone at the concentration used is unable to induce the same N-ADP release. The most likely explanation is that in the mitochondrial membrane the carrier units are distributed in two populations depending on their reactivity to CATR and BA, since it is known that these two inhibitors cannot bind simultaneously to the same carrier unit (Block et al. 1980). The two carrier populations are able to bind N-ADP, but only one of them binds CATR and exhibits a CATR-induced release of N-ADP. The AdN carriers which respond to CATR are in the CATR conformation. The other AdN carriers which bind BA and so exhibit a BA-induced release of N-ADP are in the BA conformation. The CATR of BA conformations are evidenced as they undergo the rapid release of bound N-ADP upon addition of CATR or BA.

In the absence of ADP or ATP, the transition between the two conformations is very slow, and this is illustrated by the slow phase of N-ADP release in the presence of CATR or BA. This transition is considerably accelerated by the addition of micromolar concentration amounts of ADP or ATP.

VI. Use of DGA, a Fluorescent Derivative of ATR to Probe the CATR and BA Conformations of the Membrane-Bound AdN Carrier

Dansyl-4-aminobutyryl atractyloside (DGA) is a competitive fluorescent inhibitor of ADP transport in mitochondria with Ki value close to 50 nM (Boulay et al. 1983). Upon addition of specific ligands of the AdN carrier, DGA bound to mitochondria undergoes fluorescence changes that are related to its displacement from the AdN carrier. In the experiments to be described, the fluorescence emission is measured at 520 nm through a K50 (Baltzers) filter with an excitation light centered at 340 nm. Fill the fluorescence cuvette with 2.5 ml of a medium made of 250 mM sucrose, 1 mM EDTA and 10 mM Hepes final pH 7.00 (temperature 25°C). Then add 0.2 mg of beef heart mitochondria and mix the final suspension thoroughly. Adjust the zero position of the pen recorder. Add DGA at the final concentration of 0.15 μM. Fluorescence increases up to a plateau which is calibrated to the full scale of recorder. Add 4 μl of 1 mM CATR and observe the fluorescence decrease due to the release of DGA. The effect is complete within 2 min. Repeat the same experiment with CATR added first, and control so that upon addition of DGA the final fluorescence level is the same as in the first experiment. This indicates that the fluorescence decrease is related to the specific release of bound DGA

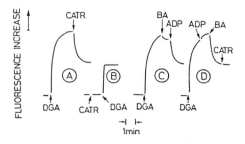

Fig. 3. Fluorescence changes induced by addition of BA, ADP, and CATR to the AdN carrier protein-DGA complex (for details see text)

upon addition of CATR. The residual fluorescent level may be attributed to DGA removed from the carrrier but still trapped within the mitochondrial membrane (Fig. 3).

Repeat the first experiment with 5 μl of 1 mM BA instead of CATR. The fluorescence decrease is slow. Then add 5 μl of 10 mM ADP. Fluorescence is rapidly quenched to the previously defined final level. As in the case of N-ADP, control the effect of ADP alone in the absence of CATR or BA: after the equilibrium has been reached in the presence of DGA and the fluorescence level has been stabilized, add 10 μl of 10 mM ADP. The fluorescence change is very limited. Then add 5 μl of 1 mM BA. Fluorescence is rapidly quenched. In a further experiment, check the effect of nontransportable natural nucleotides instead of ADP (for example, UDP, UTP, GDP, GTP, AMP, CDP, CTP).

These results can be interpreted on the same basis as those of the experiment carried out with N-ADP (Boulay et al. 1986): the AdN carrier exists in two conformations, the CATR and the BA conformations which are in equilibrium. The transition between the two conformations is slow. DGA binds to the CATR conformation. At 25°C all the carrier units rapidly take up the CATR conformation upon addition of DGA, the fluorescent signal is maximal. The bound DGA is easily removed by CATR by direct competition. On the other hand, BA is much less effective on the DGA release in the absence of ADP. However, in the presence of ADP, the equilibrium between the CATR and BA conformations is considerably accelerated. Nontransportable nucleotides have no effect on the transition between the CATR and BA conformations.

VII. Tryptophanyl Residues of the AdN Carrier as Intrinsic Fluorescent Probes

A. Principle

The isolated AdN carrier protein in detergent solution (LAPAO) exhibits a fluorescence emission spectrum characteristic of tryptophanyl groups with a broad peak centered at 330 nm, which is shifted to the red upon addition of ADP or ATP. This modification is prevented when the AdN carrier is previously incubated with CATR.

In the present study we shall observe the fluorescence changes of the isolated AdN carrier at 355 nm (coupled Corning filters 0–54 and 7–54), using an excitation light centered at 295 nm. Under these experimental conditions, the fluorescence of the AdN carrier is increased upon addition of ADP or ATP.

B. Assay

Add 0.8 ml of the purified AdN carrier in LAPAO solution, previously subjected to chromatography on AcA 202, to a quartz fluorescence cuvette containing 1.2 ml of a 136 mM glycerol solution. The cuvette is thermostated at 20°C. Record the fluorescence signal and wait for the stabilization of the base line. Adjust the recorder scale to a value equal to 25% of the signal. Then add 5 μl of a 1 mM ATP solution. The fluorescence increases rapidly, the increase corresponding to about 5% of the initial fluorescence. Within a period of 2 to 3 min, add 5 μl of 1 mM CATR. The fluorescence signal is slowly reversed to the initial value. Repeat the experiment, adding CATR prior to ATP. Observe that CATR completely prevents the fluorescence increase, which means that the CATR effect is a specific one.

Repeat the experiment described above, starting the incubation of the AdN carrier in the presence of 5 μl of 1 mM BA. Upon addition of 5 μl of 1 mM CATR. Check that CATR reverses the enhancing effect of BA and that the kinetics of the CATR-induced quenching depends on the concentration of added ADP. Control the absence of effect of nontransportable nucleotides.

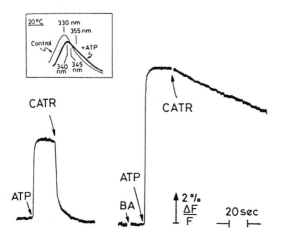

Fig. 4. Increase of the intrinsic fluorescence upon addition of ATP (or ADP). Enhancement of the fluorescence increase by BA. Reversal by CATR (for details see text)

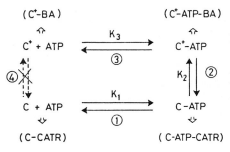

Fig. 5. Scheme illustrating the transition from conformation C (supposed to react with CATR) to conformation C^+ (supposed to react with BA). ATP (or ADP) is required for the transition between C and C^+ to occur. The direct transition from C to C^+ is forbidden

C. Interpretation

The low and high fluorescence levels assumed by the AdN carrier correspond to two different conformations of the carrier, one stabilized by CATR, the other by BA (Brandolin et al. 1985). The transition between the two conformations depends on the binding of transportable nucleotides, i.e., ADP or ATP. The intermediate level of fluorescence might correspond to a mixture of two populations of the carrier, one in the CATR conformation, the other in the BA conformation (see scheme in Fig. 5). An alternative interpretation which cannot be excluded is that a third transient conformation corresponding to the nucleotide-carrier complex may arise, the fluorescence level of which would intermediate between those of the CATR and BA conformations. The two hypothesis are not mutually exclusive.

References

Aquila H, Misra D, Eulitz M, Klingenberg M (1982) Complete amino acid sequence of the ADP/ATP carrier protein from beef heart mitochondria. Hoppe-Seyler's Z Physiol Chem 363:345–349

Block MR, Vignais PV (1984) Substrate-site interactions in the membrane-bound adenine-nucleotide carrier as disclosed by ADP and ATP analogs. Biochim Biophys Acta 767:369–376

Block MR, Pougeois R, Vignais PV (1980) Chemical radiolabeling of carboxyatractyloside by |14C| acetic anhydride. Binding properties of |14C| acetylcarboxyatractyloside ro the mitochondrial ADP/ATP carrier. FEBS Lett 117:335–340

Block MR, Lauquin GJM, Vignais PV (1982) Interaction of 3'-0-(1-naphthoyl)adenosine-5'-diphosphate, a fluorescent adenosine 5'-diphosphate analogue, with the ADP/ATP carrier protein in the mitochondrial membrane. Biochemistry 21:5451–5457

Block MR, Lauquin GJM, Vignais PV (1983) Use of 3'-0-(1-naphthoyl)adenosine 5'-diphosphate to probe distinct conformational states of membrane-bound ADP/ATP carrier. Biochemistry 22:2202–2208

Boulay F, Brandolin G, Lauquin GJM, Vignais PV (1983) Synthesis and properties of fluorescent derivatives of atractyloside as potential probes of the mitochondrial ADP/ATP carrier protein. Anal Biochem 128:323–330

Boulay F, Brandolin G, Vignais PV (1986) 6'-0-dansyl-γ-aminobutyryl atractyloside, a fluorescent probe of the ADP/ATP carrier: exploration of conformational changes of the membrane bound ADP/ATP carrier elicited by substrates and inhibitors. Biochem Biophys Res Commun 134:266–271

Brandolin G, Doussiere J, Gulik A, Gulik-Krzywicki T, Lauquin GJM, Vignais PV (1980) Kinetic binding and ultrastructural properties of the beef heart adenine nucleotide carrier protein after incorporation into phospholipid vesicles. Biochim Biophys Acta 592:592–614

Brandolin G, Dupont Y, Vignais PV (1981) Substrate-induced fluorescence changes of the isolated ADP/ATP carrier protein in solution. Biochem Biophys Res Commun 98:28–35

Brandolin G, Dupont Y, Vignais PV (1982) Exploration of the nucleotide binding sites of the isolated ADP/ATP carrier protein from beef heart mitochondria. Probing of the nucleotide sites by formycin-triphosphate, a fluorescent transportable analogue of ATP. Biochemistry 24:6348–6353

Brandolin G, Dupont Y, Vignais PV (1985) Substrate-induced modifications of the intrinsic fluorescence of the isolated adenine nucleotide carrier protein: demonstration of distinct conformational states. Biochemistry 24:1991–1997

Dupont Y, Brandolin G, Vignais PV (1982) Exploration of the nucleotide sites of the isolated ADP/ATP carrier protein from beef heart mitochondria. Probing of the nucleotide sites by naphthoyl-ATP, a fluorescent non transportable analogue of ATP. Biochemistry 21:6343–6347

Gottikh BP, Kraysesky AA, Tarussova NB, Purygin PP, Tsilevitch TL (1970) The general synthetic route to aminoacid esters of nucleotides and nucleoside-5'-triphosphates and some properties of these compounds. Tetrahedron 26:4419–4433

Guillory RJ (1979) Applications of the photoaffinity technique to the study of active sites for energy transduction. Curr Top Bioenerg 9:267–414

Klingenberg M (1981) The ADP-ATP translocation system of mitochondria. In: Lee CP, Schatz G, Dallner G (eds) Mitochondria and microsomes. Addison-Wesley, New York, pp 293–316

Lauquin GJM, Vignais PV (1976) Interaction of |3H| bongkrekic acid with the mitochondrial adenine nucleotide translocator. Biochemistry 15:2316–2322

Lijmback GWM, Cox MC, Berends W (1970) Elucidation of the chemical structure of bongkrekic acid. Isolation purification and properties of bongkrekic acid. Tetrahedron 26:5993–5999

Smith AL (1967) Preparation, properties and conditions for assay of mitochondria: slaughterhouse material, small-scale. Methods Enzymol 10:81–86

Vignais PV, Block MR, Boulay F, Brandolin G, Lauquin GJM (1985) Molecular aspects of structure-function relationships in mitochondrial adenine nucleotide carrier. In: Benga G (ed) Structure and properties of cell membranes, vol II. Molecular basis of selected transport systems. CRC Press, Boca Raton, Florida, pp 139–179

Spin Labeling of Membranes and Membrane Proteins

CLEMENS BROGER, REINHARD BOLLI and ANGELO AZZI

I. Introduction

Electron spin resonance (ESR), or electron paramagnetic resonance (EPR), has been a powerful tool during the last 20 years to derive structural and dynamic information about membranes and membrane components.

The source of an ESR signal is an unpaired electron, which may exist intrinsically in a metalloprotein (heme, iron-sulfur cluster) or be introduced artificially as a stable free radical (spin label).

The most commonly used spin labels are nitroxides of the general formula shown in Fig. 1, where the unpaired electron is located in the p_z orbital of the nitrogen atom. R_1 and R_2 are groups which give the spin label its specific reactivity and/or physicochemical properties.

The aim of this article is to introduce the spin labeling techniqeue by carrying out several basic, short experiments. There is no special equipment necessary except for an ESR spectrometer, which can be of the simplest type available.

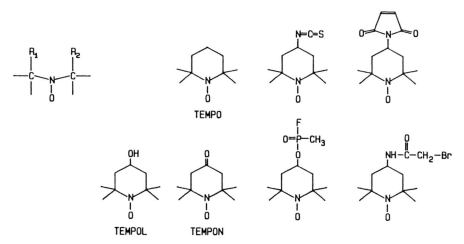

Fig. 1. General formula of a nitroxide spin label and some examples of spin labels often used in biological studies

Membrane Proteins, ed. by Azzi
©Springer-Verlag Berlin Heidelberg 1986

The shape of the ESR spectrum of a spin label depends, for example, on the concentration and the motion of the molecule and on the polarity of its environment. By attaching spin labels to proteins or lipids, the motion and lateral distribution of these molecules and their mutual interactions in a membrane can be probed.

II. Theory

The present article gives only a very short overview. For more details see Berliner (1976/1979), Knowles et al. (1976) or Wertz and Bolton (1972).

A. The ESR Signal

According to quantum mechanics, the electron possesses a spin quantum number of 1/2. This implies that the electron can assume only two spin states with components of its spin angular momentum S along a certain direction of

$$M_S = \pm 1/2. \tag{1}$$

The charge of the electron gives rise to a magnetic moment μ. The magnetic moment is related to the spin angular momentum by the equation:

$$\mu = g\beta S. \tag{2}$$

If an external magnetic field is applied, the electron can assume only two states, with components of the magnetic moment along the direction of the field (z-axis) of:

$$\mu_z = -g\beta M_S, \tag{3}$$

where g is a quantum mechanical constant, the so-called g-value, which for a free electron is 2.0023. For nitroxide spin labels, similar g-values are observed. The g-values of transition metals, on the other hand, can be in the range from about 1 to 10.

β is the Bohr magneton ($9.274 \ 10^{-21}$ erg/gauss), which is defined as follows:

$$\beta = eh/2 \ cm, \tag{4}$$

where e = electron charge, h = Planck constant, c = light velocity, m = electron mass.

The energy E of a magnetic dipole in a magnetic field is defined as

$$E = \mu \cdot H, \tag{5}$$

where H = magnetic field strength.

For the electron we substitute Eq. (1) and (3) into Eq. (5) to obtain the energies of the two allowed states (Zeeman energies):

$$E_{1,2} = \pm \, g\beta H/2 \tag{6}$$

and the energy required for the transition between the two states is:

$$\Delta E = g\beta H. \tag{7}$$

This transition can be induced by irradiation with electromagnetic waves of the energy hv, where v is the frequency of the wave, thus the condiditon for resonance is:

$$hv = g\beta H. \tag{8}$$

In spin label experiments the frequency v is usually held constant at about 9.4 GHz (microwaves) and the magnetic field strength is varied. The transition will then occur at about 3300 Gauss.

In a nitroxide spin label the unpaired electron is located in the p_z orbital of the nitrogen atom, and is thus within close distance of the nitrogen nucleus. The most abundant isotope ^{14}N has a nuclear spin quantum number of I=1. Therefore $2 \cdot I + 1 = 3$ spin states with the magnetic quantum numbers $M_I = -1, 0$ and +1 are allowed for the nitrogen nucleus. The magnetic moment of the electron interacts with the nuclear moment. The latter produces a small magnetic field at the position of the electron either along the applied magnetic field ($M_I=1$) or opposite to it ($M_I=-1$). In the case of $M_I=0$ no additional magnetic field is produced. Thus for a nitroxide spin label the ESR signal is split into three lines (hyperfine splitting). In the case of a fast isotropically moving spin label the three lines are symmetrical and since the probabilities for all three states of the nuclear spin are the same the lines have the same intensities, as shown in Fig. 2.

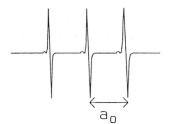

Fig. 2. Spectrum of a nitroxide spin label moving fast and isotropically. The isotropic hyperfine splitting constant is a_o

The hyperfine splitting constant a as well as the g-value are slightly dependent on the polarity of the solvent. If a spin label partitions between water and a lipid phase, the high field peak is resolved into two components due to the differences of the g- and a-values in the two phases as shown in Fig. 3.

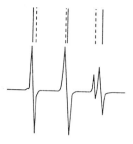

Fig. 3. Partition of TEMPO between a polar and a hydrophobic phase. *Top* solid lines represent the spectrum of TEMPO in water and the dotted lines the spectrum of TEMPO in the lipid phase. *Bottom* superimposition of the two spectra

B. Spectral Anisotropy

If the g-values and the hyperfine splitting constants are measured in crystals of spin labels, the values obtained are different for different orientations of the spin label molecules with respect to the external magnetic field. This phenomenon is called spectral anisotropy (Fig. 4).

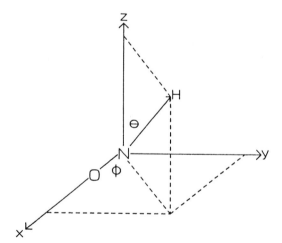

Fig. 4. Molecular coordinate system for nitroxide spin labels. The x-axis lies along the N−O bond, the z-axis along the p_z oribital of nitrogen. The y-axis is perpendicular to the x- and z-axes. H indicates the direction of the external magnetic field. θ is the angle between the magnetic field and the z-axis. ϕ is the angle between the x-axis and the projection of H in the x-y-plane

If the values are known for the orientation of the external field along the three main axes of the spin label, the values for the hyperfine splitting and for g can be calculated according to the following formulas (approximations):

$$g = g_{xx} \sin^2 (\theta) \cos^2 (\phi) + g_{yy} \sin^2 (\theta) \sin^2 (\phi) + g_{zz} \cos^2 (\phi) \qquad (9)$$

and

$$a = a_{xx} \sin^2 (\theta) \cos^2 (\phi) + a_{yy} \sin^2 (\theta) \sin^2 (\phi) + a_{zz} \cos^2 (\phi). \qquad (10)$$

Fig. 5. Powder spectrum of a nitroxide spin label

For nitroxide spin labels, the values measured along the main axes are approximately

$$g_{xx} = 2.0089, g_{yy} = 2.0069 \text{ and } g_{zz} = 2.0027 \qquad (11)$$

$$a_{xx} = a_{yy} = 6 \text{ gauss and } a_{zz} = 32 \text{ gauss.} \qquad (12)$$

For rapid isotropically moving spin labels, the orientations are averaged and thus the measured values are:

$$g_0 = 1/3 (g_{xx} + g_{yy} + g_{zz}) \qquad (13)$$

$$a_0 = 1/3 (a_{xx} + a_{yy} + a_{zz}) \qquad (14)$$

Another extreme situation is met when the orientation of the spin labels is random but the solution is frozen (rigid glass spectrum or powder spectrum). The spectrum is the sum of the spectra of spin labels in all orientations and is shown in Fig. 5.

Intermediate isotropic tumbling velocities of spin labels lead to spectra between the powder spectrum and the free spectrum. With common ESR techniques rotational correlation times (t_c) between 10^{-7} s (powder spectrum) and 10^{-11} s (free spectrum) can be measured. This is the range of correlation times which is observed for small molecules like lipids. Proteins move mare slowly and will practically always show a nearly rigid glass spectrum. In order to study their motions, other techniques (e.g., saturation transfer EPR) have to be applied.

Totational correlation times can be calculated by the Debye equation for Brownian rotational diffusion:

$$t_c = \frac{4\pi r^3 \eta}{3kT}, \qquad (15)$$

With η = viscosity, r = radius of the molecule, k = Boltzmann constant, T = absolute temperature, or it may be estimated from ESR spectra by using a semi-empirical formula derived by Stone et al. (1965)

$$t_c = 6.5 \ 10^{-10} \cdot \Delta H_0 \cdot (\sqrt{h_0/h_{-1}} - 1) \text{ s,} \qquad (16)$$

where ΔH_0 = linewidth of the central line in gauss, h_0, h_{-1} = intensities of the central and the high field lines, respectively.

The validity range of this formula is from about 10^{-9} to 10^{-11} s.

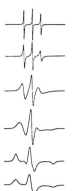

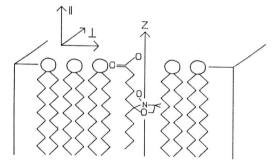

Fig. 6. Spectra of spin-labels rotating isotropically. The rotational correlation time is increasing from top to bottom from about 10^{-11} to 10^{-7} s

Fig. 7. Spin labeled fatty acid incorporated into a bilayer

If the spin label is not moving isotropically, but rotates about one axis preferentially, the ESR spectrum shows a different shape. Such anisotropic motion is found, for example, when a spin labeled fatty acid or a phospholipid is incorporated into a lipid bilayer.

The preferential rotation occurs around the hydrocarbon chain (bilayer normal), which coincides with the spin label z-axis. The motion in the x-y plane is restricted. Thus, maximal and minimal hyperfine splittings are observed, when the applied magnetic field is parallel or perpendicular to the bilayer normal ($a_\parallel$ and $a_\perp$), respectively.

If the motion is completely restricted to the rotation around the z-axis, $a_\parallel$ reaches its highest value and becomes a_{zz} and $a_\perp$ becomes minimal and equal to $a_{xx}=a_{yy}$. If the motion does not occur exclusively around the main axis of the fatty acid, $a_\parallel$ and $a_\perp$ are resolved and can be measured from the spectrum. A practical measure for the anisotropy of the motion is the order parameter:

$$S = (a_\parallel - a_\perp) / (a_{zz} - 1/2(a_{xx} + a_{yy}) \cdot a_0/a_0' , \qquad (17)$$

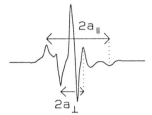

Fig. 8. Spectrum of a spinlabel undergoing anisotropic motion (rotation preferentially around the z-axis)

where a_0/a_0' is a correction factor for the different polarity of the crystal environment where a_{xx}, a_{yy} and a_{zz} are measured, as compared to the bilayer where $a_{\parallel}$ and $a_{\perp}$ are measured.

$$a_0/a_0' = (a_{xx}+a_{yy}+a_{zz}) / (a_{\parallel} + 2a_{\perp})$$

S = 0 : isotropic motion
S = 1 : rotation exclusively around the main axis of the fatty acid

The order parameter can be used to determine the mean angular deviation of the spin label from the bilayer normal.

$$S = 1/2 \cdot (\cos\beta + \cos^2\beta) \tag{18}$$

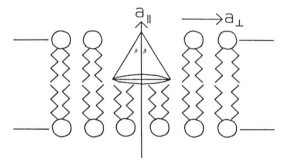

Fig. 9. Motion of a lipid spin label in a membrane restricted to a cone with an opening angle of 2β

III. Experimental Procedures

The following experiments are described:
1. Dependence of the ESR spectrum on the concentration of a spin label
2. Determination of the melting point of a lipid
3. Anisotropic motion of a fatty acid at different depths in the bilayer
4. Spin labeling of the mitochondrial phosphate translocator

A. Equipment and Chemicals

— ESR spectrometer (e.g., Varian E 104 A)
— Capillaries
— Eppendorf centrifuge
— Ultrasonic bath
— Pasteur pipets

Chemicals for experiment 1:
— 4-Oxo-2,2,6,6-tetramethylpiperidino-1-oxy (TEMPON); 25 mg ml^{-1} in water

Chemicals for experiment 2:
— Dimyristoyl phosphatidylcholine; 100 mg ml^{-1} in chloroform
— 2,2,6,6-tetramethylpiperidino-1-oxy (TEMPO); 1.5 mg ml^{-1} in water

Chemicals for experiment 3:
— Spin labeled stearic acids (ALDRICH) having the spin labels in the positions C_5, C_{12} and C_{16}; 1 mg ml^{-1} in ethanol
— Dipalmitoyl phosphatidyl choline (DPPC); 100 mg ml^{-1} in chloroform

Chemicals for experiment 4:
— beef heart mitochondria in buffer 1; 20 mg protein ml^{-1}
— maleimide spin labels (MSL, ALDRICH) (see Fig. 10); 5 mg ml^{-1} in ethanol
— cysteine (MERCK); 200 mM, neutralized with KOH
— Bovine serum albumin (BSA); 10% in water
— mersalyl; 7.5 mg ml^{-1} in water
— N-ethyl maleimide (NEM); 200 mM in ethanol
— ferricyanide; 500 mM in water
— ascorbate (neutralized); 250 mM in water

Buffers:
— buffer 1: (labeling buffer) 10 mM Tris/HCl pH 7.4, 0.25 M sucrose, 1 mM EDTA
— buffer 2: (washing buffer) 10 mM Tris/HCl pH 7.4, 0.25 M sucrose, 1 mM EDTA, 0.2% BSA, 5 mM ferricyanide

B. Experiment 1: Concentration Effect

The shape of the EPR spectrum depends on the average distance between spin labels: At close distances broadening on the lines is observed due to

R	Spacer length (Å)
MSL1	–	6.8
MSL2	$CONHCH_2CH_2$	11.7
MSL3	$CONH(CH_2)_2O(CH_2)_2$	15.3

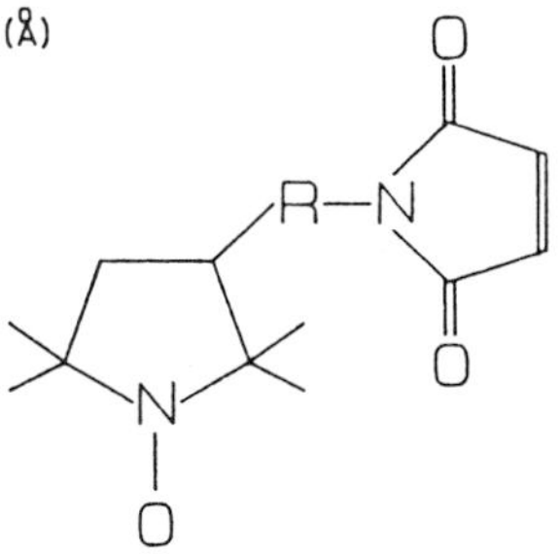

Fig. 10. Maleimide spin labels

dipolar interaction between spins. At even closer distance spin-spin exchange occurs.

Procedure

The following mixtures are prepared:
50 μl TEMPON + 250 μl water
50 μl TEMPON + 100 μl water
50 μl TEMPON + 50 μl water
50 μl TEMPON
    Measure the spectra of the solutions under the following conditions:
Microwave power 5–10 mW
Modulation frequency 100 kHz
Modulation amplitude 1 gauss
    Calculate in each case an average distance between two TEMPON molecules, assuming that each molecule is confined to a cube. It is found that the spectrum is distorted by dipolar interactions already at the lowest concentration, at which the average distance between the molecules is about 40 Å. A spin-spin exchange spectrum is observed at the highest concentration (average distance between molecules about 22 Å).

C. Experiment 2: Measurement of the Melting Point of Dimyristoyl Phosphatidylcholine (DMPC)

The partition of TEMPO between the lipid and water phase in a lipid suspension in water is temperature-dependent. Below the melting point of the lipid the spin probe is excluded from the lipid due to its crystalline structure. When the lipid melts, TEMPO becomes more soluble in the lipid. The concentration ratio of spin label in the two phases is proportional to the ratio of the peak heights of the two resolved high field peaks.

Procedure

75 μl of DMPC is put into a tube and the solvent is evaporated. 75 μl of water and 5 μl of TEMPO are added and the tube is vigorously shaken at 40°C (above the phase transition temperature of the lipid) until the suspension is homogeneous. The sample is put into a capillary and measured at different temperatures between 0° and 40°C.

Microwave power 5–10 mW
Modulation frequency 100 kHz
Modulation amplitude 0.5 gauss

Calculate the ratio R for all the temperatures. R = H/(H+P) where H and P are then peak heights of the high field ESR signal from the spin probe in the hydrophobic and the polar phase, respectively.

Plot R against the temperature. The melting point of the lipid is at the point of inflection of the curve.

The melting point reported in the literature is 23.5^{o}C. A phase transition can be observed below the melting point of the lipid at about 10°C.

D. Experiment 3: Anisotropic Motion of Spin Labeled Stearic Acid Incorporated into a Lipid Bilayer

Add 40 μl of DPPC to three tubes. To each of the tubes add 2.5 μl of one of the spin labeled stearic acids. The solvents are evaporated carefully with a stream of nitrogen. Add 250 μl of water to the tubes and sonicate until the solutions are clear. Measure the ESR spectra in capillaries at room temperature with the following settings:

Microwave power 5–10 mW
Modulation frequency 100 kHz
Modulation amplitude 5 gauss

– Calculate the order parameters [according to formula (17)] for the three fatty acids by using the following hyperfine splitting constants: a_{xx} = 5.9 G, a_{yy} = 5.4 G and a_{zz} = 32.9 G.
– Calculate the opening angles of the cones in which the spin labels move by using formula (18) and solving the quadratic equation.

Typical results are presented in the table below

Spin label	Order parameter	Angle
C_5	0.62	44 degrees
C_{12}	0.52	51 degrees
C_{16}	0.41	58 degrees

E. Experiment 4: Spin Labeling of the Mitochondrial Phosphate Translocator (Houstek et al. 1983)

The electro neutral phosphate translocator is the main transport system for phosphate in heart mitochondria (for a review see Pederson and Wehrle 1982). It is strongly inhibited by N-ethyl maleimide, which binds covalently to sulf-hydryl groups of cysteine residues of the protein. The active carrier consists probably of two identical polypeptides with a molecular weight of 34,500, of which only one is active at a time.

Procedures

1. Spectra of MSL Labeled Mitochondria. Mitochondria are labeled with the three MSL's according to the procedure described below but without pre-treatment with mersalyl and NEM. The samples are measured in capillaries at room temperature.

2. Labeling of mersalyl and NEM inhibited mitochondria. 2 ml of mitochondria are washed twice with buffer 1. Sedimentation is performed in an Eppendorf centrifuge for 5 min at 4°C. The pellet is taken up in 8 ml of buffer 1 (5 mg protein ml^{-1}). The suspension is treated as follows. Add 75 μl of mersalyl (25 nmol mg^{-1} of protein), shake and incubate for 1 min at 0°C. Add 80 μl of NEM and incubate for further 2 min at 0°C. Stop the reaction with 200 μl of cysteine and sediment the mitochondria in the Eppendorf centrifuge for 5 min. Wash the mitochondria 4 times with 12 ml buffer 1, which contains 0.5 mM cysteine during the first washing procedure. The pellet is taken up in 8 ml of buffer 1 and divided into 4 equal portions.

The portions are incubated with the three different MSL's (50 nmol mg^{-1} of protein) for 2 min at 0°C (2 x MSL 1: 24 μl, 1 x MSL 2: 32 μl and MSL 3: 36 μl. The reaction is stopped by the addition of 20 μl of cysteine. The suspensions are supplemented with 40 μl BSA and 20 μl of ferricyanide to protect the spin label against reduction. The mitochondria are washed 4 times with buffer 2, until no ESR signal is observed in the supernatants. The sediments are taken up in 0.25 ml of buffer 1 and the ESR spectra are measured.

The second sample of MSL 1 labeled mitochondria is washed twice with buffer 1 in order to remove residual ferricyanide and resuspended in 0.25 ml of the same buffer. 5 μl of ascorbate (5 mM final concentration) is added and spectra are measured every 2 to 5 min up to about 1 h.

Results and Discussion

Direct Labeling of Mitochondria. All three spin labels will give highly mobile spectra. The spin labels are attached not only to the SH-groups of the phosphate carrier but also of many other proteins. Since the side chain of cysteine is long and flexible, it is expected that the spin labels do not represent the motion of the protein as a whole, which would be near a powder spectrum.

Labeling of Pretreated Mitochondria. Mersalyl is a reversible ligand for SH-groups. During the short incubation time it reacts with the SH-groups of the phophate carrier which are known to be the most reactive of all in mitochondria. During the incubation with NEM all SH-groups except those protected by mersalyl are covalently labeled and during the washing procedure mersalyl is removed again. At this stage all the SH groups are labeled with NEM, except those of the phosphate carrier which are free again and can be reacted with the spin label maleimides.

The sample labeled with MSL 1 (shortest spacer) yields an EPR spectrum which consists of two components: an immobile and a mobile one. As the spacer length is increased, the immobile signal vanishes, whereas the fast-moving species increases.

This result is interpreted in the following way. One SH-group is located in a pocket of the protein or in the membrane where the motion of the spin label is restricted (slow component of the spectrum). The other SH-group is located at the surface of the protein and is quite mobile (fast component of the spectrum). As the spin labels with longer spacers are used the nitroxide moiety is able to reach the surface and then even stick out of the protein. From these measurements it can be estimated that the first SH group is located at a depth of about 10 $\mathring{A}$.

In the reduction experiment with ascorbate, the ratio between the heights of the two low field absorption peaks of the superimposed spectra are calculated and plotted against time. The mobile species is very sensitive toward reduction by ascorbate, whereas the immobile species is quite resistant. This means that the mobile spin label is in a hydrophilic milieu and the immobile label in a hydrophobic region where ascorbate does not penetrate.

References

Berliner LJ (1976/1979) Spin labeling: Theory and applications, vol I, 1976; vol II, 1979. Academic Press, London New York

Houstek J, Bertoli E, Stipani I, Pavelka S, Megli FM, Palmieri F (1983) FEBS Lett 154: 185–190

Knowles PF, Marsh D, Rattle HWE (1976) Magnetic resonance of biomolecules. Wiley, London

Pederson PL, Wehrle JP (1982) In: Martonosi AN (ed) Membranes and transport, vol I. Plenum Press, New York London, pp 645–663

Shimshick EJ, McConnell HM (1973) Biochemistry 12:2351–2360

Stone TJ, Buckman T, Nordio PL, McConnell HM (1965) Proc Natl Acad Sci USA 54: 1010–1017

Wertz JE, Bolton JR (1972) Electron spin resonance, elementary theory and practical applications. McGraw-Hill, New York

IV. Protein Reconstitution

Functional Reconstitution of the Mitochondrial Cytochrome b-c$_1$ Complex: Effect of Cholesterol

M.J. NAŁĘCZ and A. AZZI

I. Introduction

The mitochondrial cytochrome b-c$_1$ complex is an oligomeric lipoprotein catalyzing the electron transfer from ubiquinol (QH$_2$) to oxidized cytochrome c. The enzyme contains as electron carriers two spectroscopically distinguished species of cytochrome b, an iron-sulfur protein with a binuclear iron-sulfur cluster and cytochrome c$_1$. The electron transport activity of the b-c$_1$ complex is coupled to vectorial proton translocation, and it is now well established that for each electron pair passing through the enzyme, four protons appear on the cytoplasmic side of the inner mitochondrial membrane. Two of these protons are uncoupler-insensitive and result from the overall oxidation of QH$_2$ by cytochrome c, the other two are uncoupler-sensitive and are translocated from the matrix side of the inner mitochondrial membrane. The latter is indicated by the extrusion of only two positive charges per electron pair. The mechanism by which b-c$_1$ complex translocates protons through the membrane is still a matter of debate, and different hypotheses have been proposed to explain this phenomenon (see Nałęcz 1986 for review). At variance with cytochrome c oxidase, where both a vectorial electron and proton translocation are responsible for coupling, in the case of the b-c$_1$ complex only a proton translocation occurs.

Whatever the mechanism, the vectorial proton translocation and the chemical proton ejection associated with the redox reaction lead to acidification of external medium which can be measured by means of different techniques. Knowing the rate of electron flow through the system, one may calculate the proton to electron ratio (H$^+$/e$^-$ratio), an important parameter characterizing functionally active mitochondrial b-c$_1$ complex.

The vectorial proton transport leads also to the formation of the transmembrane potential, negative inside the mitochondria. The potential buildup slows down the electron flow which is then utilized only to maintain the energized state of the membrane (state 4 respiration). In the presence of an uncoupler (protonophor) and valinomycin (potassium ionophor) the electrochemical gradient of protons is dissipated (H$^+$/e$^-$ratio reflects only the appearance of the chemical proton) and the electron flow reaches its highest level (state 3 respiration). The extent of electron flow stimulation (state 3 versus state 4) is

Membrane Proteins, ed. by Azzi
© Springer-Verlag Berlin Heidelberg 1986

known as the coupling or the respiratory control of the system (see Nicholls 1982).

Isolated mitochondrial cytochrome b-c$_1$ complex can be reconstituted into phospholipid vesicles using different procedures, depending mainly on the type of detergent associated with the enzyme (see Eytan 1982 and Casey 1984 for reviews). Reconstituted b-c$_1$ vesicles show respiratory control and H$^+$/e$^-$ ratios similar to those usually obtained in mitochondria. The importance of this fact is clear: it enables the use of the purifiedprotein complex to study its proton-translocating function without any interference from other membrane components. Although proteoliposomes are not free from disadvantages (see, e.g., Casey 1984) they still serve as the most convenient model system.

One of the basic conditions for successful reconstitution of respiratory chain complexes is the tight coupling of proteoliposomes. This can be achieved by lowering unspecific membrane permeability for ions (especially protons) and providing an optimal phospholipid surrounding (composition, fluidity) required for high enzymatic activity of the studied protein. In the present experiment, cholesterol will be used as factor modulating coupling of the reconstituted b-c$_1$ complex.

A detailed account of cholesterol influence on membrane structure and functions is beyond the scope of this introduction; the reader is referred to the following reviews (Demel and de Kruijff 1976, Razin and Rottem 1978, Quinn 1981, Davis 1983, Yeagle 1985). The following, however, is a brief summary of the aspects relevant to the experiment described below.

The hydrophobic character of the cholesterol molecule makes it highly insoluble in water, but it allows its easy association with membranes, at a 1:1 and even higher molar ratio with respect to the phospholipids. Cholesterol is buried in the lipid core of the membrane and does not extend into the polar head region of the bilayer. The majority of its rigid ring structure is in direct contact with phospholipid hydrocarbon chains (thus influencing their motion), whereas its hydroxyl head group forms a hydrogen bond with the ester carbonyl of the phospholipids (thus influencing the structure of the lipid/water interface). Cholesterol has a wide variety of effects on the physical properties of membranes. The most dramatic is the cholesterol-induced change in the enthalpy and cooperativity of the gel to liquid-crystalline phase transition in phospholipid bilayers: below the phase transition temperature cholesterol induces disorder in the lipid acyl chains, whereas above this temperature it has a solidifying effect. Cholesterol also influences the permeability of membranes to various small molecules. It is now well established that increasing content of cholesterol in the membrane decreases its permeability to sugars, ions, and small hydrophobic substances. A similar effect is also noted on the electrical conductivity and the water permeability of the membrane. Furthermore, cholesterol influences membrane proteins. Reports have been published that suggest a role for cholesterol in the state of aggregation of some proteins (e.g., band 3 from erythrocytes), direct protein-cholesterol interaction (e.g., erythrocyte glycophorin) and modulation of the activity of many membrane-bound enzymes and transporting systems (see Yeagle 1985 for review). Interestingly, cholesterol was shown to affect protein isolated from membranes, like the inner mitochondrial membrane, which do not contain it normally (see, e.g., Kramer 1982, for the cholesterol-induced activation of the ADP/ATP translocator).

In the present experiment the reconstitution of the b-c₁ complex will be performed using a variation of a direct-incorporation method (see Eytan 1982). The aim of the present experiment is to show the effect of cholesterol on the coupling and the proton translocating activity of the b-c₁ proteoliposomes.

II. Equipment, Materials, and Reagents

A. Equipment

— Refrigerated MSE-65 Superspeed centrifuge with a 12 x 10 fixed angle rotor, or equivalent
— Amino DW-2a spectrophotometer fitted with a magnetic stirring device and a thermostatically controlled cuvet holder
— Radiometer PHM 64 pH meter with an Ingold LoT 405-M3 combined pH electrode
— W+W 600 chart recorder
— MSE sonicator or equivalent, equipped with a microtip
— Vortex mixer
— Small Sephadex G-50 columns (Pasteur pipets filled with a pre-swollen resin)
— Round-bottom thick glass tubes (10 ml), with stoppers
— Small glass balls

B. Material

— Mitochondrial cytochrome b-c₁ complex is isolated from beef heart mitochondria according to a standard cholate-solubilization, salt precipitation procedure (Rieske 1967) and stored at −80°C. The usual preparation contains 6–8 nmol heme b and 3–4 nmol heme c₁ per mg of protein when estimated spectrophotometrically using the following extinction coefficients: $25.6 \text{ mM}^{-1}\text{cm}^{-1}$ (562–577 nm) for heme b and $20.1 \text{ mM}^{-1}\text{cm}^{-1}$ (553–540 nm) for heme c₁.

C. Chemicals

— Asolectin (phospholipid extract from soybeans) is a commercial product
  from Associated Concentrates and is used without any further purification
— Cholesterol is from Fluka
— Sephadex G-50 Coarse is from Pharmacia
— Cytochrome c (type VI from horse heart) is from Sigma Chemicals Co.
  Its stock solution is prepared in water at about 1 mM. The exact concentration is estimated spectrophotometrically using the extinction coefficient
  of 19.4 $mM^{-1}cm^{-1}$ at 550–540 nm
— 2-methyl-3-undecyl-1,4-naphtoquinol (UNH, the electron donor for the
  b-c_1 complex) is a commercial product from the Alfred Bader Library of
  Rare Chemicals, Division of Aldrich Chemical Company
— Antimycin A is from Sigma Chemical Co. and is prepared as 1 mg ml^{-1}
  solution in ethanol. Its exact concentration is estimated spectrophotometrically using the extinction coefficient of 4.8 $mM^{-1}cm^{-1}$ at 320 nm
— Valinomycin and carbonylcyanide-m-chlorophenylhydrazone (CCCP) are
  from Sigma Chemical Co. and are used either as a mixed solution in ethanol
  (0.2 mM: 0.25 mM, valinomycin: CCCP) or separately, at the same concentration
— All other chemicals are of the highest purity commercially available

III. Experimental Procedure

A. Reconstitution of the b-c_1 Complex

Lipid samples used for reconstitution of the mitochondrial cytochrome b-c_1
complex are prepared with increasing content of cholesterol according to the
following protocol:
1. 55 mg asolectin
2. 55 mg asolectin + 1.6 mg cholesterol (5 mol%)
3. 55 mg asolectin + 3.4 mg cholesterol (10 mol%)
4. 55 mg asolectin + 5.4 mg cholesterol (15 mol%)
5. 55 mg asolectin + 7.6 mg cholesterol (20 mol%)
6. 55 mg asolectin + 13.0 mg cholesterol (30 mol%)

All samples are prepared in 10 ml round-bottom thick glass tubes with
stoppers. Lipids (+ cholesterol) are dissolved in 1 ml chloroform: methanol
(1:1, v/v) and dried under a stream of nitrogen to give a thin film on the bottom of the tube. When lipids are dry, 1 ml of 200 mM sodium phosphate buffer (pH 7.4) is added to each sample. All tubes are filled with the nitrogen gas

and closed with stoppers. Lipids are allowed to swell (about 1 h, room temperature), being occasionally vortexed. Small glass beads may be added to speed up the formation of multilamellar vesicles during mixing. When the samples become translucent, the stock preparation of the isolated b-c$_1$ complex is added to each suspension (about 15 nmol of cytochrome b per sample). The tubes are again filled with the nitrogen gas, closed and placed in the refrigerator for about 1 h. After this time each sample is sonicated (with cooling, under a stream of nitrogen) for 2 min (50% duty, at 6 μ peak to peak amplitude) with the MSE sonicator, followed by centrifugation at 35,000 g for 1/2 h. The opalescent supernatants are filtered through small Sephadex G-50 columns equilibrated with 10 mM choline chloride-10 mM KCl solution. The final concentration of the enzyme in proteoliposomes is measured spectrophotometrically as described above (section Material).

B. Reduction of Quinol Derivative

UNH is reduced in the following way: A carefully weighed amount (5–10 mg) of the compound is placed in a glass tube and dissolved in 400 μl of a mixture of DMSO and ethanol, 1:1 (v/v). To the resulting bright yellow solution about 50 μl of 2 M sodium borohydride is added stepwise in 10 μl portions until the liquid becomes reddish brown. At this point the solution is acidified to eliminate the excess of borohydride by stepwise additions of 10 μl portions of 2 M HCl until no visible evolution of gas occurs. The resulting solution of reduced UNH is colorless and is kept under nitrogen. Usually, the final concentration is about 40 mM and can be used directly for studies on coupling of the b-c$_1$ proteoliposomes. For the measurements of proton appearance, however, the UNH solution must not contain any excess of HCl. In this case the reduced form of UNH is extracted from the above solution by two subsequent washings with 1 ml of cyclohexane. The cyclohexane phase is collected, dried under a stream of nitrogen and the resulting yellowish powder is weighed and re-dissolved in the ethanol/DMSO mixture.

C. Measurement of Coupling of Proteoliposomes

As mentioned in the Introduction, coupling of proteoliposomes is estimated from the extent of stimulation of electron flow through the b-c$_1$ complex induced by a protonophor/ionophor mixture added to the energized system. The electron transport activity of the enzyme is measured spectrophotometrically as the antimycin A-sensitive reduction of cytochrome c by quinol. Reduction of cytochrome c is monitored as the change in absorbance at 550–540 nm and the measurement is performed as follows:

1. Fill a spectrophotometer cuvet with 1.5 ml of 100 mM potassium phosphate buffer (pH 7.4) and supplement the medium with cytochrome c (final concentration of about 20 μM). Add the reconstituted b-c$_1$ proteoliposomes to give
the final concentration of about 0.2 μM heme b and preincubate the sample
for 1 min in order to reach the desired temperature. The cuvette chamber is
thermostated at 25°C and the sample is continuously mixed with a small magnetic bar. Start the reaction by addition of the reduced form of UNH (final
concentration of about 20 μM) and record the initial rate of cytochrome c reduction.

2. Repeat the measurement in the presence of antimycin A (final concentration of about 1 μM), added after proteoliposomes.

3. For the uncoupled rate of cytochrome c reduction, repeat the first measurement in the presence of valinomycin + CCCP (add 10 μl of the stock mixture, after proteoliposomes).

4. Repeat the measurement (3) in the presence of antimycin A. The reaction rate, reflecting nonenzymatic reduction of cytochrome c by UNH, should
be the same as for sample (2).

Both rates measured in samples (1) and (3) have to be diminished by the
antimycin A-insensitive reaction. The obtained values of enzymatic reduction
of cytochrome c are used to calculate the ratio between the electron flow
through the reconstituted b-c$_1$ complex in the presence and absence of the uncoupler, respectively. The final value represents coupling (respiratory control)
of proteoliposomes.

The above procedure is repeated for each sample of the reconstituted enzyme (proteoliposomes prepared with different amounts of cholesterol).

D. Measurement of H^+/e^- Ratio

In this part of the experiment, changes of pH are measured simultaneously
with cytochrome c reduction. This is achieved by inserting the thin Ingold pH
electrode directly into the spectrophotometric cuvette. Care must be taken
not to interfere with the light path of the spectrophotometer. Since the measurement is performed with an open chamber, it is important to work with
dimmed light when the photomultiplier is operating (recording of the absorbance changes).

In order to measure the acidification of the external medium, accompany-
ing enzymatic reduction of cytochrome c by quinol, a nonbuffered medium
must be used (75 mM choline chloride, 25 mM KCl). The rapid production of
a large transmembrane electrical potential due to the uncompensated proton
extrusion would tend to inhibit further H^+ translocation. Valinomycin, in the
presence of external K^+, allows collapse of the charge differential and, conse-

quently, more extensive and detectable proton translocation. Addition of
CCCP to this system produces a rapid equilibration of the proton gradient by
increasing the permeability of the vesicles to protons, and thus makes the vectorial H^+ translocation undetectable.

The measurement is performed as follows:

1. Fill a spectrophotometer cuvet with 1.5 ml of 75 mM choline chloride,
25 mM KCl, add b-c₁ proteoliposomes at the final concentration of about
0.2 μM heme b and supplement the solution with cytochrome c (final concentration of about 10 μM) and valinomycin (final concentration of about 1 μM).
Adjust the pH of the sample to about 7.0 (on the recorder scale, previously
calibrated to this pH) with diluted solutions of either HCl or NaOH. When the
pH is stable and of desired value, start the reaction by the addition of UNH
(final concentration of about 10 μM). Both rates of the pH change and cytochrome c reduction are recorded simultaneously. When the reaction is over,
calibrate the pH change with 4–5 subsequent additions of 5 μl of freshly prepared 1 mM oxalic acid. The latter procedure is vital to know exactly how
many proton equivalents appeared in the medium during the reaction. It has
to be performed for each sample separately, since the buffering capacity may
differ slightly from one sample to another.

2. Repeat the measurement in the presence of CCP (final concentration of
2.5 μM).

3. Repeat the measurement in the presence of antimycin A (final concentration of 1 μM).

The same procedure is applied to all preparations of the b-c₁ proteoliposomes reconstituted with different amounts of cholesterol.

Only the initial rates of both proton appearance and electron flow are
taken for calculation of the H^+/e^- ratio. The number of protons appearing
in the medium is estimated directly from the calibration with oxalic acid. The
number of electrons then transported, being equal to the number of reduced
cytochrome c molecules, is estimated spectrophotometrically from the measured absorbance change and the extinction coefficient of the cytochrome.

The pH measurement performed in the presence of antimycin allows the
correction for a possible artifact due to the addition of quinol as well as for
the nonenzymatic reduction of cytochrome c. Therefore the number of protons released and the rate of electron transport measured in this experiment
should be subtracted from those obtained with the noninhibited enzyme.

The protons released in the presence of CCCP are generated in the chemical oxidation of the quinol, while those appearing in the absence of CCCP
are generated both by the chemical reaction and by the vectorial translocation.

IV. Results

Results of the present experiment can be summarized in the table:

mol% Cholesterol	Respiratory control	H^+/e^- Ratio	
		+valinomycin	+valinomycin+CCCP
None	3.8	1.71	0.89
5	4.6	1.84	0.91
10	7.3	1.98	0.92
15	4.4	1.80	0.99
20	3.2	1.42	0.95
30	1.6	1.05	0.90

The sample in which the highest values of the respiratory control (coupling) and the H^+/e^- ratio (+valinomycin) are obtained represents the optimal reconstitution system. For the b-c_1 complex studied as described above the optimum is usually found at 10 mol% cholesterol.

V. Comments

One of the problems in studies of the b-c_1 complex is the lack of a commercially available, good electron donor. UNH described in the present experiment is by far not the ideal substrate for the enzyme, undergoing a relatively high autooxidation and a nonenzymatic reaction with cytochrome c. For more detailed and accurate studies another quinone derivative, 2,3-dimethoxy-5-methyl-6-decyl-1,4-benzoquinone (DBH), should be used. Although this highly stable analog of ubiquinone-2 (its reduced form can be stored for weeks in $-80^{\circ}C$ without any substantial loss of reactivity) is not commercially available to date, it can be synthesized and reduced as described (Wan et al. 1975).

   A regulatory role of cholesterol on the b-c_1 complex in vivo is out of discussion, since this sterol is absent from the inner mitochondrial membrane, where the enzyme is naturally located. However, the fact that cholesterol modulates the reconstitution of the complex, probably as a consequence of its effect on physical parameters of the phospholipid bilayer, may be helpful in studies with the purified enzyme.

References

Casey RP (1984) Biochim Biophys Acta 768:319–347
Davis JH (1983) Biochim Biophys Acta 737:117–171
Demel RA, Kruijff B de (1976) Biochim Biophys Acta 457:109–132
Eytan GD (1982) Biochim Biophys Acta 694:185–202
Kramer R (1982) Biochim Biophys Acta 693:296–304
Nałęcz MJ (1986) J Bioenerg Biomembr 118:21–38
Nicholls DG (1982) Bioenergetics. An introduction to the chemiosmotic theory. Academic Press, London New York
Quinn PJ (1981) Prog Biophys Mol Biol 38:1–104
Razin S, Rottem S (1978) Trends Biochem Sci 3:51–55
Rieske JS (1967) Methods Enzymol 10:239–245
Wan YP, Williams RH, Folkers K, Leung KH, Racker E (1975) Biochem Biophys Res Commun 63:11–15
Yeagle PL (1985) Biochim Biophys Acta 822:267–287

Changes of the Membrane Surface Potential Measured by Amphiphilic Fluorescent and ESR Probes

M.J. NAŁĘCZ, A. SZEWCZYK, and L. WOJTCZAK

I. Introduction

A difference in the electric potential is usually formed at the boundary between two phases due to the transfer of ions and/or electrons from one phase to another. This occurs when ions or electrons dissociate from one phase and become associated with molecules of the other phase or when some mobile ions have different solubility (mobility) in either of the phases. The potential difference between the surface of phase A and a point infinitely distant from the interface within phase B is defined as the surface potential (more precisely: potential at the surface) of phase A.

Biological and artificial membranes usually carry fixed electric charges on their sufaces. The most common negatively charged groups are phosphate monoesters (phosphorylated proteins), phosphate diesters (phospholipids) and carboxylic groups (proteins, glycoproteins, phosphatidylserine), whereas main positively charged groups are protonated amino groups (proteins, phosphatidylethanolamine), quaternary ammonium cations (phospholipids) and guanidines (proteins). In most natural membranes, the resulting net charge is negative at neutral pH. Fixed charges are the origin of the surface potential extending from the membrane surface over a distance of several nanometers into the aqueous phase. The surface potential, in turn, produces an uneven distribution of ions in the medium around the membrane. The concentration of ions bearing the charge opposite to that of the membrane (i.e., cations in case of negatively charged membranes) is increased in the immediate vicinity of the membrane surface, whereas the concentration of ions of the same sign (i.e., anions) is decreased as compared to the concentration of these ions in the bulk solution. The relation between the surface charge density, the surface potential, and the ionic strength of the medium is described by the Gouy-Chapman theory (Overbeek 1952, see also Wojtczak and Nałęcz 1985 for review).

The surface charge density of biological and artificial membranes can be altered under experimental conditions by several factors, e.g., pH (changing protonation of ionizable polar groups), divalent and polyvalent cations (able to complex phospholipid head groups), and ionic amphiphilic compounds

Membrane Proteins, ed. by Azzi

(locating their hydrophobic moieties in the lipid core of the membrane and protruding charged hydrophilic groups outside the membrane surface). The latter way of changing the membrane surface charge will be exemplified in the present experiment by using small nonsolubilizing amounts of anionic and cationic detergents, although many natural substances, as long-chain fatty acids and their CoA thioesters, aliphatic amines, certain polypeptides and charged phospholipids produce similar effects (see also Wojtczak and Nałęcz 1979, Nałęcz et al. 1980).

Since the membrane surface potential governs the distribution of ions within the aqueous phase in the vicinity of the membrane, it can control the activity of membrane-bound enzymes interacting with ionized substrates, activators and/or inhibitors as well as control the transmembrane ion fluxes (for review see Wojtczak and Nałęcz 1985).

The surface potential can be determined by means of several techniques. These include membrane electrophoresis revealing the so-called zeta potential, i.e., the potential at the shear layer of an adhering film of the medium moving together with the membrane (see, e.g. Sherbet 1978), methods based on changing properties of membrane-bound probes, e.g., 4-heptadecyl-umbelliferone or other membrane-bound pH indicators (Fromherz and Masters 1974, Vaz et al. 1978), and methods based on the affinity of probes to the membrane. The latter include measurements with the use of either fluorescent (Azzi 1975) or electron spin resonance (Mehlhorn and Packer 1979) probes. Applicability of amphiphiles as probes of the surface potential requires that their partitioning between the membrane and the aqueous medium be easily and accurately determined.

A. Fluorescent Probes

The general requirement of an easy and accurate measurement is fulfilled by several fluorescent compounds, e.g., 8-anilino-1-naphthalene sulfonate (ANS). 6-toluidyl-1-naphthalene sulfonate and ethidium bromide, whose fluorescence spectrum and/or quantum yield depend on the hydrophobicity of the environment (Azzi 1975). In the case of ANS, the anionic probe used in the present experiment, a blue shift of the emission maximum and a considerable increase of the quantum yield upon transfer of the probe from aqueous into a hydrophobic medium (Fig. 1) make it possible to neglect its fluorescence in water and to ascribe all measured fluorescence practically to the probe bound to the membrane (Haynes 1974). The dependence of ANS binding on the salt concentration is compatible with the Gouy-Chapman theory and enables the determination of the surface potential and hence the surface charge density (Gibrat et al. 1983). In a much simpler way, ANS can be used to estimate re-

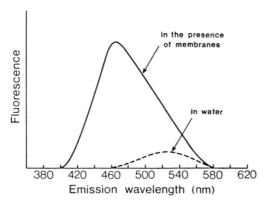

Fig. 1. Emission spectrum of ANS in water and in the presence of microsomes

lative values of the surface potential and, in particular, changes of this potential (Wojtczak and Nałęcz 1979, 1985). Caution must be taken, however, when working with energizable membranes. In the presence of the transmembrane potential, the ANS fluorescence response becomes complex, depending on both the surface and the transmembrane potentials (Robertson and Rottenberg 1983).

B. Spin-Labeled Probes

Another group of surface potential probes based on affinity changes includes charged amphiphilic compounds containing a paramagnetic moiety in their hydrophilic part. Since the ESR signal of such compounds is altered when they are bound to membranes, as shown in Fig. 2 for the signal of 4-[N,N-dimethyl-N-dodecyl]-ammonium-2,2,6,6-tetramethylpiperidine-1-oxyl (CAT_{12}) used in the present experiment, their partitioning between the aqueous phase and the membrane can be determined and used as a measure of the membrane surface potential (Mehlhorn and Packer 1979). A change of the surface potential can thus be determined directly from the spectra taken at two different membrane states or conditions. However, there are also problems with ESR probes. First, the paramagnetic nitroxide group can be reduced by certain enzymes present in natural membranes, as has been shown for mitochondria, and thus addition of some oxidants to the medium is recommended (Hashimoto et al. 1984). Second, the formation of a transmembrane potential in energizable membranes promotes the transmembrane migration of the probe and makes it unsuitable for surface potential measurements (Hashimoto et al. 1984, Wojtczak and Szewczyk 1986).

In the present experiment, both ANS and CAT_{12} are used to determine changes of the surface potential of rat liver microsomes (endoplasmic reticulum membranes).

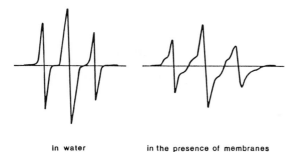

in water   in the presence of membranes

Fig. 2. Electron spin resonance spectrum of CAT_{12} taken in water and in the presence of microsomes

II. Equipment, Materials and Reagents

A. Equipment

— For isolation of microsomes: Potter homogenizer (25 ml), preparative centrifuge (e.g., Sorvall with rotor SS-34) and ultracentrifuge (e.g., Beckman L-5 with rotor Ti60).

— For ANS fluorescence measurements: Spectrofluorimeter, e.g., Perkin-Elmer MPF-3L, set up at 366 nm excitation and 460 nm emission wavelengths, with No. 39 filter. Both slits are about 8 and sensitivity at 30 or 100.

— For CAT_{12} electron spin resonance measurements: ESR spectrometer, e.g., Varian E-104A, with quartz capillaries. The instrument is set up at 100 kHz modulation frequency, 1 gauss modulation amplitude and 5–10 mW microwave power.

B. Material

— Fresh rat liver is chopped into small parts, suspended in the medium of 250 mM sucrose and 10 mM Tris-HCl (pH 7.4) and homogenized with a Potter homogenizer. The suspension is centrifuged at 600 g for 10 min (sedimentation of cells and tissue fragments). The pellet is discarded and the supernatant is centrifuged at 15,000 g for 15 min. The pellet is discarded or used to obtain mitochondria for other purposes, and the supernatant is centrifuged for 60 min at 100,000 g to sediment microsomes. Microsomal pellet is suspended in 250 mM sucrose/10 mM Tris-HCl (pH 7.4).

C. Chemicals

- ANS is a commercial product and is used from the stock solution of 5 mM in water.
- CAT_{12} (synthesized according to Mehlhorn and Packer 1979), 1 mM in water.
- Cetyltrimethylammonium bromide (CTAB), the cationic detergent (commercial), 10 mM in water.
- Sodium dodecylsulfate (SDS), the anionic detergent (commercial), 10 mM in water.
- Sucrose and Tris are of highest purity commercially available.

III. Experimental Procedures

The medium used for all measurements is 250 mM sucrose/10 mM Tris-HCl (pH 7.4).

A. Measurements of ANS fluorescence

Fill a fluorimeter cuvet with 2.5 ml of the medium and add microsomal suspension corresponding to 1−2 mg protein. Read the fluorescence and, if any present, suppress it to zero on the recorder scale. Make the following traces:

1st trace (control): Titrate the particles with 5 μl portions of ANS, recording the increasing fluorescence of the probe. When a subsequent addition fails to produce a measurable increase of the fluorescence (the membranes become saturated with the dye), stop the titration.

2nd trace (+CTAB): Add 10 μl of CTAB (final concentration about 40 μM), stir the suspension and start the titration as before.

3rd trace (+SDS): Add 10 μl of SDS (final concentration about 40 μM) and repeat the titration with ANS.

Calculations

Partitioning of the charged amphiphilic compound between the membrane and aqueous phase is affected by the surface potential, because the local concentration of the compound in the immediate vicinity of the membrane is different than that in the bulk solution. Such an effect is described by the Boltzmann distribution:

$$C_o = C_{bulk} \exp (-zF\ \psi_s/RT), \tag{1}$$

where C_O is the concentration of an ion in the vicinity of the membrane surface, C_{bulk} denotes the concentration in the bulk solution, z is a positive or negative integer denoting the number of charges of the ionized compound, F is the Faraday constant (96,487 C mol^{-1} or 23,063 cal volt^{-1} mol^{-1}), R is the gas constant (8.3143 J mol^{-1} K^{-1} or 1.987 cal mol^{-1} K^{-1}), T is the absolute temperature and ψ_s is the potential at the membrane surface.

Based on the Boltzmann distribution, the following dependence of the dissociation constant of an ion (in this case the charged amphiphile) upon the surface potential can be formulated:

$$K'_d = K^o_d \exp(-zF\psi_s/RT), \tag{2}$$

where K^o_d is the dissociation constant of the compound to the uncharged membrane (the so-called "intrinsic dissociation constant"), K_d' is the apparent dissociation constant to the membrane bearing a surface potential ψ_s and other symbols are as before. This equation can be used to calculate the absolute surface potential if both K^o_d and K'_d can be measured (see Gibrat et al. 1983). In a much simpler way, ANS may be used to estimate *changes* of the surface potential according to the formula:

$$\Delta\psi_s = \frac{RT}{zF} \ln \frac{K''_d}{K'_d}, \tag{3}$$

where $\Delta\psi_s$ is a difference, or a change, of the surface potential between two states of the membrane ($\Delta\psi_s = \psi_s'' - \psi_s'$), and K_d'' and K_d' are corresponding apparent dissociation constants.

In the present experiment $\Delta\psi_s$ can be calculated in the following way:
— Prepare double reciprocal plots of ANS fluorescence (expressed in arbitrary units) versus the probe concentration for all three traces of membrane titration (e.g., Fig. 3).
— Read the values of dissociation constants for ANS in control microsomes, microsomes treated with CTAB and treated with SDS.
— Use these values for calculation of the membrane surface potential changes ($\Delta\psi_s$) according to formula (3). The value of the integer z for ANS is −1.

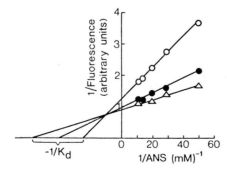

Fig. 3. Double reciprocal plot of ANS fluorescence versus ANS concentration. Titration of rat liver microsomes. *Closed circles* control membranes; *open circles* microsomes treated with SDS; *triangles* microsomes treated with CTAB. Experimental conditions as described

B. Measurements of ESR signal of CAT$_{12}$

Fill a spectrometer capillary with the sucrose-Tris suspension of microsomes containing $1-2$ mg protein ml^{-1} and CAT$_{12}$ in the final concentration of $30 \mu M$ ($15 \mu l$ of 1 mM CAT$_{12}$ added to 0.5 ml of the suspension). Set the spectrometer as described above and measure the ESR spectrum for control microsomes.

Repeat the measurement first in the presence of added CTAB, and second in the presence of SDS (both in the final concentration of about $40 \mu M$, i.e., $2 \mu l$ of 10 mM stock solution added to 0.5 ml of the microsomal suspension).

Calculations

Partition coefficient P (water vs. membrane) can be determined from the spectrum of CAT$_{12}$ as examplified in Fig. 4 (see also Quintanilha and Packer 1977, Mehlhorn and Packer 1979). P values thus obtained for control microsomes, microsomes treated with CTAB and microsomes treated with SDS can subsequently be used for calculation of the change of the membrane surface potential, as described for the fluorescence measurements with ANS. The following formula has to be used:

$$\Delta\psi_s = \frac{RT}{zF} \ln \frac{P'}{P''} \quad , \tag{3}$$

where P' and P'' are partition coefficients for two different membrane conditions. The value of z equals +1 for CAT$_{12}$.

The values of changes of the microsomal surface potential calculated from both types of experiment (ANS fluorescence and ESR spectrum of CAT$_{12}$) should be similar.

IV. Comments

It has to be remembered that the surface potential as measured by membrane probes reflects a random distribution of charges on the membrane surface. If charges are not uniformly spread over the membrane, the potential sensed by membrane enzymes and carriers may differ from that measured as described

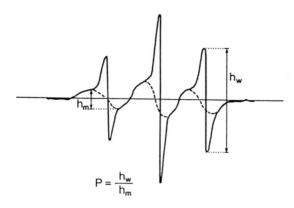

Fig. 4. Electron spin resonance spectrum of CAT_{12} in the presence of rat liver microsomes. The way of estimating of the partition coefficient of $CAT_{12}(P)$ is presented according to Quintanilha and Packer (1977)

$$P = \frac{h_w}{h_m}$$

here. Nevertheless, changes of the random surface potential have been found to explain variations in the activity of several membrane-bound enzymes and transporting systems (Wojtczak and Nałęcz 1985).

References

Azzi A (1975) Q Rev Biophys 8:237−316
Fromherz P, Masters B (1974) Biochim Biophys Acta 356:270−278
Gibrat R, Romien C, Grignon C (1983) Biochim Biophys Acta 736:196−202
Hashimoto K, Angiolillo P, Rottenberg H (1984) Biochim Biophys Acta 764:55−62
Haynes DH (1974) J Membr Biol 17:341−366
Mehlhorn RJ, Packer L (1979) Methods Enzymol 56:515−526
Nałęcz,MJ, Zborowski J, Famulski KS, Wojtczak L (1980) Eur J Biochem 112:75−80
Overbeek JThG (1952) In: Kruyt HR (ed) Colloid Science, vol I, chap 4. Elsevier, Amsterdam
Quintanilha AT, Packer L (1977) FEBS Lett 78:161−165
Robertson DE, Rottenberg H (1983) J Biol Chem 258:11039−11048
Sherbet GV (1978) The Biophysical Characterisation of Cell Surface, Academ Press, London New York
Vaz WLC, Nicksch A, Jahning F (1978) Eur J Biochem 83:299−305
Wojtczak L, Nałęcz MJ (1979) Eur J Biochem 94:99−107
Wojtczak L, Nałęcz MJ (1985) In: Benga G (ed) Structure and Properties of Cell Membranes, vol II. CRC Press, Boca Ration,Florida, pp 215−242
Wojtczak L, Szewczyk A (1986) Biochem Biophys Res Commun 136:941−946

Reconstitution of Cytochrome c Oxidase

N. CAPITANIO and S. PAPA

I. Introduction

The reconstitution of coupled electron transport in cytochrome c oxidase is realized by removal of cholate from a starting suspension of purified complex (prepared as described by Errede et al. 1978) and a mixture of phospholipids. The reconstituted system so obtained (proteoliposomes) is constituted by small unilamellar vesicles with a diameter of 500–600 $\mathring{A}$ (Casey et al. 1979, Racker 1979, Nicholls 1981).

The orientation in the membrane of the asymmetrically shaped complex is not random; the amount of cytochrome c oxidase with externally exposed cytochrome c binding sites represents 80% of the total. The oxidation rate of ferrocytochrome c by cytochrome c oxidase vesicles is enhanced some tenfold in the presence of FCCP, whose protonophoric activity collapses transmembrane proton gradients (Fig. 1). The stimulation of respiration by the uncoupler has been equated with the phenomenon of respiratory control in mitochondria, and indicates that the energy of oxidation of cytochrome c is conserved in the form of a transmembrane proton gradient.

The oxidation of ferrocytochrome c is accompanied by an ejection of protons into the suspending medium (Fig. 2).

Valinomycin is added to the system to allow potassium counterflow to compensate for the charge imbalance created by the net inward movement of negative charges.

When, in these conditions, the vesicles are pulsed with ferrocytochrome c, there is an immediate acidification of the medium followed by a slow decay to a new plateau value.

When the number of the induced turnovers is relatively low (Casey et al. 1979, Racker 1979), the number of protons released is almost equal to the amount of ferrocytochrome c oxidized. In the presence of FCCP, the same pulse of ferrocytochrome c results in a net alkalinization of the suspension.

The ratios H^+/e^- can be calculated by comparing the amount of protons released or consumed at the maximum extent with the amount of reducing equivalents added.

Membrane Proteins, ed. by Azzi
©Springer-Verlag Berlin Heidelberg 1986

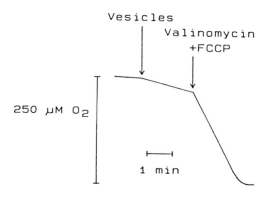

Fig. 1. Respiratory control index of oxidase-containing phospholipid vesicles. Oxygen consumption was measured in a closed cuvet, using an oxygen electrode. For the medium composition see "Experimental Conditions". The reaction was started by adding lipid vesicles containing 0.03 μM heme a+a$_3$ and the activation of respiration was induced by 1 μg ml^{-1} valinomycin and 3 μM FCCP

II. Equipment and Reagents

A. Equipment

- Sonicator (Mod. W185F, Heatsystem Ultrasonic, Inc)
- O_2-Polarograph (Yellow Spring, Ohio, USA)
- Potentiometric recorder (Perkin-Elmer)
- Spectrophotometer (Perkin-Elmer Lambda UV-VIS)
- Electrometer (Keithly Inst. Cleveland, Ohio, USA, Mod 604)
- pH-Combination electrode (Beckman N. 39505)

B. Reagents

- Tris
- Hepes
- EDTA
- KCl
- Cytochrome c (Sigma Type VI)
- Cytochrome c oxidase
- FCCP or CCCP
- Choline-Cl
- Valinomycin

III. Experimental Procedure

A. Preparation of Cytochrome Oxidase Vesicles (Cholate-Dialysis Procedure)

— 40 mg ml^{-1} purified soybean phospholipids (asolectin) in: Na-cholate 1.5% 100 mM K-Hepes [4-(2-hydroxyethyl)-1-piperazineethanesulfonic acid] pH 7.5
— sonication in a Sonicator cell disruptor (Mod. W185F, Heatsystem Ultrasonic, Inc) under a flux of nitrogen (three exposures, at approx. 60 W, for 2 min, with 2 min intervals)
— addition of purified bovine heart cytochrome c oxidase to give a final concentration of 6.0 μM on a heme basis,
— dialysis against: I)   100 vol. of: 100 mM K-Hepes, pH 7.2 for 4 h
                    II)   200 vol. of: 10 mM K-Hepes
                                          27 mM KCl
                                          73 mM sucrose, pH 7.2 for 4 h
                    III) 200 vol. of: 1 mM K-Hepes
                                          30 mM KCl
                                          79 mM sucrose, pH 7.2 overnight

All solutions are adjusted to the indicated pH with KOH. The procedure is carried out at 4°C.

B. Measurement of Respiratory Control Index (Potentiometric Assay)

Oxygen consumption is measured potentiometrically in a thermostatically controlled (25°C) all-glass cell, equipped with a magnetic stirring device and a Clark oxygen electrode (4004 YSI, Yellow Spring, Ohio, USA) coated with a high sensitivity membrane (YSI 5776). The output of the electrometer is connected to a Perkin-Elmer strip-chart recorder.

C. Experimental Conditions

Medium composition:	Activation of coupled	Activation of uncoupled
40 mM KCl	respiration by the	respiration by the addition
10 mM K-Hepes	addition of vesicles:	of:
0.1 mM K-EDTA	0.03 μM heme a+a$_3$	1 μg ml^{-1} valinomycin plus
25 mM K-ascorbate		3 μM FCCP (or CCCP)
50 μM cytochrome c		
pH 7.4 (dioxygen con-		
centration 250 μM)		
see Fig. 1		

D. Ferrocytochrome c Preparation

- 80 mg of cytochrome c (SIGMA, type VI) in 3 ml: 100 mM K-Hepes pH 7.4 40 mM Tris
- reduction by addition of a few grains of Na-dithionite;
- dialysis against: I) 1.1. 100 mM Choline-Cl for 4 h
                     II) 1.1 100 mM Choline-Cl overnight.

E. Ferrocytochrome c Assay

Measurement of ferrocytochrome c is carried out spectrophotometrically, at room temperature with a Perin-Elmer, LAMDA UV/VIS spectrophotometer using an ϵ_{550} RED-OX = 21.0 mM^{-1}. The cytochrome c is oxidized using a few grains of ferricyanide.

F. Measurement of Redox-Linked Proton Ejection from Cytochrome c Oxidase Vesicles

In reductant pulse experiments the oxidized resting (Michell and Moyle 1983, Brunori et al. 1979) oxidase is supplemented, in the aerobic state, with ferrocytochrome c. The H$^+$/e$^-$ ratio is measured from the extent of electron transfer (given by the amount of added ferrocytochrome c). In the oxidant pulse experiments a lightly buffered suspension of cytochrome oxidase vesicles is made anaerobic and respiration is induced by injecting a small, known quantity of air-saturated H$_2$O (cytochrome c is maintained reduced by an excess of ascorbate-TMPD). Figure 2 shows proton translocation in a reductant pulse experiment. The vesicles are incubated in a thermostatically controlled (25°C) glass cell, equipped with a magnetic stirring device. Proton translocation is measured with a pH combination electrode (Beckman N. 39505) connected to an electrometer amplifier (Mod. 604 Keithley Instruments, Cleveland, Ohio, USA) and a strip-chart fast responding potentiometric recorder. The system gives a precision of 10^{-3} pH unit, an overall response time (10–90% change) of 0.4 s.

G. Experimental Conditions

- Cytochrome oxidase vesicles, 1 μM hemes a+a$_3$ in:
  100 mM Choline-Cl
  0.2 mM Choline-Hepes

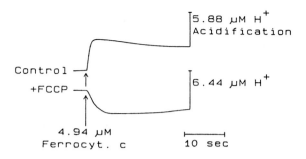

Fig. 2. H$^+$ ejection by oxidase-containing phospholipid vesicles. For the composition of the buffer, see "Experimental Conditions". The reaction was started by adding 4.9 μM ferrocytochrome c. The acidification was followed by using a pH electrode. The amount of cytochrome c oxidase was 1 μM. *Lower trace* represents the proton uptake by the system when supplemented with 3 μM FCCP. Appropriate amounts of acid were added to calibrate the system

0.1 mM Choline-EDTA
5 mM K-Cl
1 μg ml^{-1} valinomycin pH 7.0;
— pulse of 4–6 μM ferrocytochrome c, adjusted to the same pH of the suspension with a precision of 0.005 unit pH. The same assay can be carried out in the presence of 3 μM FCCP (or CCCP). The redox linked pH change are quantitated with small aliquots of a standard solution of 10 mM HCl.

References

Brunori M, Colosimo A, Rainoni O, Wilson MT, Antonini E (1979) J Biol Chem 254: 10769–10775
Casey RP, Chappel JB, Azzi A (1979) Biochem J 182:149–156
Errede B, Kamen MD, Hatefi Y (1978) Methods Enzymol 55:614–627
Michell P, Moyle J (1983) FEBS Lett 151:167–178
Nicholls P (1981) Int Rev Cytol Suppl 12:327–388
Racker E (1979) Methods Enzymol 55:699–711

Measurement of Pyruvate Transport in Mitochondria

K.A. NAŁĘCZ and A. AZZI

I. Introduction

Mitochondria are surrounded by two membranes. The only protein known so far to modify the permeability of the outer membrane is porin. Its activity resembles that of channel (Roos et al. 1982) with an exclusion limit of about 5,000. Due to porin, the outer membrane is permeable to ions and metabolites. Several carrier systems have instead been detected in the inner mitochondrial membrane. Transport or exchange systems for phosphate, 2-oxo-monocarboxylates, glutamate,adenine nucleotides, glutamate/aspartate, dicarboxylates, tricarboxylates, 2-oxoglutarate have been demonstrated (for review see LaNoue and Schoolwerth 1979). In mitochondria the primary transport of substrates is driven by a proton gradient or a transmembrane electrical potential built up by the respiratory chain.

The first technique used to demonstrate the existence of different carriers was based on mitochondrial swelling. Since radioactive substrates have been made available, tracer techniques have allowed direct measurements of the transport. In order to quantitate the transport, one needs to stop the reaction immediately after a chosen incubation time. This can be achieved by a potent, quickly reacting inhibitor. If such an inhibitor is not available, it is necessary to separate the mitochondria from the external medium as fast as possible. To do so two techniques are mostly used, either membrane filtration or centrifugation through silicone oil, the so-called centrifugal filtration (Palmieri and Klingenberg 1979). Whatever method is applied, some water which cannot be ascribed to the matrix volume remains associated with the mitochondrial fraction. This water is either located in the intermembrane space or belongs to the external medium, which has not been removed completely. The volume of these two compartments can be estimated by measuring, in a parallel experiment, the amount of radioactive sucrose remaining associated with the mitochondria after centrifugation or filtration. Sucrose does not penetrate the inner mitochondrial membrane and, therefore, from the amount of radioactivity in the mitochondrial fraction one can calculate the volume of the water space outside the matrix.

Membrane Proteins, ed. by Azzi
© Springer-Verlag Berlin Heidelberg 1986

In the following experiment the transport activity of the carrier is measured by using pyruvate as a substrate. This transport protein is specific for pyruvate, phenylpyruvate, 2-oxobutyrate, 2-oxoisocaproate and other 2-oxomonocarboxylates (Paradies and Papa 1976, 1977, Halestrap 1975, Nałęcz et al. 1984). The carrier shows, at least in vitro, two activities, the uptake of monocarboxylates (antiport with OH⁻ or symport with protons) and the electroneutral exchange of monocarboxylic anions. Both reactions are specifically inhibited by 2-cyanocinnamic acid and (some of) its derivatives. The mechanism of inhibition by these substances has been intensively studied (Halestrap 1975, 1978, Halestrap et al. 1980).

According to the chemiosmotic theory the transport of electrons along the respiratory chain is coupled to the ejection of protons into the cytosol. This reaction creates an electrical potential (negative inside) and a pH gradient (alkaline inside) across the inner mitochondrial membrane (Boyer et al. 1977). The overall protonmotive force Δp is defined by the relationship:

$$\Delta p = \frac{\Delta \mu_{H^+}}{F} = \Delta \psi - 2.3 \; \frac{RT}{F} \cdot \; \Delta pH , \qquad (1)$$

where R is the gas constant, T the absolute temperature, F the Faraday constant, $\Delta \psi$ the transmembrane electrical potential and $\Delta \mu_{H^+}$ the electrochemical proton gradient.

Several methods exist for measuring ΔpH and $\Delta \psi$. In this experiment the distribution of a weak acid (acetic acid) was used for the measurement of ΔpH (Nicholls 1977). $\Delta \psi$ can be measured from the distribution of either inorganic ions, such as Rb^+ or K^+, or certain lipophilic cations, e.g., tetraphenylphosphonium (TPP^+) or triphenylmethylphosphonium ($TPMP^+$) (Bakeeva et al. 1970).

II. Principle of the Experiment

The experiment is performed with freshly prepared rat liver mitochondria. The uptake of [1–^{14}C] pyruvate is measured at low temperature in order to slow down its metabolism (Halestrap 1978). Moreover, the choice of pyruvate labeled in position 1 excludes that decarboxylation products of the substrate are also measured.

The time course of pyruvate uptake is measured with and without 1 mM 2-cyano-4-hydroxycinnamate. In order to determine the volume of the mitochondrial matrix, it is assumed that both membranes are permeable to water, while sucrose penetrates only the outer membrane and therefore is present in the water trapped between the mitochondria and in the intermembrane space.

By comparison of the distribution of acetate and TPMP$^+$ and the one of pyruvate the mechanism of pyruvate uptake (according to ΔpH or $\Delta\psi$) can be inferred. Using the same data, the protonmotive force can be calculated according to Eq. (1).

III. Equipment and Reagents

A. Equipment

— High speed centrifuge
— UV/VIS spectrophotometer
— Liquid scintillation counter
— 65 scintillation vials (20 ml)
— Eppendorf centrifuge, type 3200 with 50 tubes (1.5 ml)
— Cryostat (6°C) or a cold-room
— Water bath (100°C)
— Motor-driven Potter homogenizer with Teflon pestle (50 ml)
— Water pump
— Hamilton syringe for 50 μl
— Conical tubes for 10 ml (about 50)
— Set of adjustable automatic pipets

B. Reagents

1. Isolation medium: 225 mM mannitol, 75 mM sucrose, 0.5 mM EGTA, 3 mM Tris-HCl, pH 7.4 (500 ml).
2. Biuret solution: 1.5 g CuSO$_4$ and 6 g Na-K-tartrate is dissolved in 500 ml water, 300 ml of 10% NaOH is added and the mixture is supplemented with water to 1 liter.
3. Deoxycholate (sodium salt), 10% w/v solution in water (100 μl per experiment).
4. Perchloric acid, 14% solution in water (5 ml).
5. Wacker silicone oil mixture: AR200/AR20 mixed 3:1 (25 ml)
6. Scintillation cocktail: 4.9 g of PPO (2.5-diphenyloxazole) and 0.21 g of POPOP 2,2'-p-phenylen-bis(5-phenyloxazol) are dissolved in 700 ml of toluene, 300 ml of 98.8% ethanol is added.
7. Incubation medium: 150 mM sucrose, 1 mM MgCl$_2$, 0.5 mM EDTA, 9 μM

antimycin, 12 μM rotenon, 2.5 μg ml^{-1} oligomycin, 30 mM Tris-HCl, pH 7.4.

8. 5 mM [1–^{14}C] pyruvic acid, sodium salt (Amersham) in 10 mM Tris-HCl, pH 7.4, 5 Ci mol^{-1} (2 ml).

9. 0.1 M 2-cyano-4-hydroxycinnamate (Sigma): 18.9 mg of this cinnamate derivative is dissolved in about 1 ml of H$_2$O, sonicated and neutralized with KOH. The exact concentration can be calculated from the extinction measurement at 340 nm (ϵ = 22 M^{-1}cm^{-1}, Halestrap and Denton 1975).

10. 7.5 mM [^{14}C] acetic acid, sodium salt (Amersham), 3 Ci mol^{-1} (200 μl).

11. 8 μM [^{3}H] Triphenylmethylphosphonium bromide, TPMP$^+$ (Amersham), 25 μCi ml^{-1} (25 μl).

12. 225 mM [^{14}C] sucrose (Amersham), 5 μCi ml^{-1} (50 μl).

13. ^{3}H$_2$O (Amersham), 50 μCi ml^{-1} (50 μl).

IV. Experimental Procedure

A. Preparation of Rat Liver Mitochondria

Male Wistar rats are used. The liver is removed from a rat killed by decapitation and immediately put into a beaker containing 10–15 ml of ice-cold isolation medium (reagent 1). The liver is chopped finely with scissors and the liquid decanted. This washing procedure is repeated with reagent 1, until blood is no longer visible. Small portions of the tissue are homogenized in a Potter homogenizer. The total volume of reagent no 1 used for this step should be about 50 ml per liver. The homogenate is then centrifuged at 600 g for 5 min in order to remove whole cells and nuclei. The supernatant is collected and centrifuged at 10,000 g for 10 min. The supernatant is discarded, while the pellet is gently resuspended in reagent 1 (about 30 ml), homogenized again and centrifuged at 600 g for 3 min. The supernatant is decanted and centrifuged at 10,000 g for 10 min. The pellet of mitochondria is resuspended in reagent 1, homogenized and sedimented at 10,000 g for 10 min. This washing procedure should be repeated twice. Finally, the mitochondria are resuspended in 2 ml of reagent 1 and, after protein estimation, used for the experiments.

B. Protein Estimation

A modified biuret method is applied (Gornall et al. 1949). Into 3 test tubes (one blank and two samples) the following reagents are pipeted:

- 1.50 ml water (1.55 ml for blank)
- 1.45 ml biuret solution (reagent 2)
- 30 μl 10% deoxycholate (reagent 3)
- 50 μl sample (mitochondria)

The content of the tubes is mixed, heated for 60 s at 100°C, and thereafter immediately cooled in ice. The absorption of both samples (A_1 and A_2) is read at 540 nm against the blank. The protein content (in mg ml^{-1}) is calculated according to the following formula:

$$(A_1 + A_2) \times 100 \quad [\text{mg protein ml}^{-1}] \tag{2}$$

C. The Technique of Separation of Mitochondria from the External Medium

The uptake of pyruvate and the distribution of [^{14}C] acetate and [^{3}H] TPMP$^+$ are measured by applying the technique of centrifugal filtration through silicone oil.

36 Eppendorf tubes (20 for part D, 15 for part E, 1 for equilibration) are filled in the following way: 100 μl of 14% perchloric acid (reagent 4) is put at the bottom and 600 μl of the silicone oil mixture (reagent 5) is layered on the acid. After the incubation of the mitochondria with the substrate as described in parts D and E 500 μl of mitochondrial suspension is put on top of the oil layer and centrifuged for 2 min. Since the time of centrifugation and stopping of the centrifuge is rather long, it is recommended to treat the samples one by one.

After centrifugation, 50 μl of the upper aqueous phase is removed with a Hamilton syringe (where indicated), put into a scintillation vial and 5 ml of scintillation cocktail (reagent 6) is added. In all samples the remaining upper aqueous layer is removed by suction with a water pump. Subsequently, about 200 μl of water is carefully added onto the oil surface and removed by suction. After this washing step, 50 μl of the fraction below the oil is carefully removed with a Hamilton syringe and the radioactivity is counted (5 min per sample). The total radioactivities in the layers below and above oil are obtained by multiplying the measured values by 2 and 10, respectively.

DPM (disintegrations per min) are read from the scintillation counter (or calculated from the measured counts min^{-1}(cpm) and the counting efficiency). For double labeled samples a program for ^{3}H+^{14}C DPM is applied, which best discriminates the two isotopes. For the Beckman LS3801 instrument, for example, channel 1 (lower limit 0, upper limit 400) and channel 2 (lower limit 400, upper limit 670) are used. For single labeled samples only channel 1 is used, with 0 as the lower limit. The upper limit is set at 400 for tritium and at 670 for ^{14}C.

In order to facilitate further calculations, the specific radioactivity (in DPM $nmol^{-1}$) of all labeled compounds should be established before the experiment by measuring the radioacitvity of 5 μl of the isotope solutions (in triplicates).

D. The Time Course of Pyruvate Uptake by Mitochondria

500 μl of incubation medium (reagent 7) is pipeted to each of 20 conical tubes and equilibrated at $6^{o}C$. Mitochondria (1.1 mg protein, 10–20 μl) are added to 500 μl of the medium and incubated for 30 s. The uptake reaction is started with 50 μl of 5 mM $[1-^{14}C]$ pyruvate (reagent 8). The final concentration of pyruvate, after tenfold dilution, is 0.5 mM. After 10 s, 20 s, 40 s, 1 min, 1.5 min, 2 min, 2.5 min, 3 min, 4 min, 5 min, samples of 500 μl are removed and layered on the oil and treated as described in part C.

Pyruvate uptake is followed also in the presence of 1 mM 2-cyano-4-hydroxycinnamate. In this case, together with the mitochondria, 5 μl of the inhibitor (reagent 9) are added. 30 s incubation in the presence of this compound are sufficient for complete inhibition (Halestrap et al. 1980). 50 μl of 5 mM $[1-^{14}C]$ pyruvate are added and the samples are incubated and processed as described above.

For this part of the experiment it is sufficient to take only 50 μl samples from below the oil for radioactivity measurements. The uptake of $[1-^{14}C]$ pyruvate (in nmol mg^{-1} of mitochondrial protein) is plotted as a function of time, both in the absence and presence of the inhibitor. The 2-cyano-4-hydroxycinnamate-sensitive transport is calculated as the difference and plotted versus time.

E. Mechanism of Pyruvate Transport

In order to establish the mechanism of pyruvate transport, one has to measure the distribution of pyruvate, acetate, and $TPMP^{+}$ across the inner mitochondrial membrane.

Incubation conditions are the same as in part D. 1.1 mg of mitochondrial protein are preincubated for 30 s and the reaction is subsequently started with the labeled substance according to the following scheme:

Tube number	Incubation medium (μl)	2-cyano-4-hydroxy cinnamate	Labeled substance	Volume of substance (μl)
1–3	500	Absent	[^{14}C]Pyruvate	50
4–6	500	Present	[^{14}C]Pyruvate	50
7–9	500	Absent	[^{14}C]Acetate	50
10–12	545	Absent	[^{3}H]TPMP$^+$	5
13–15	530	Absent	^{3}H$_2$O + [^{14}C]Sucrose[a]	10 10

[a]Tritiated water is added after 30 s of preincubation, whilst sucrose is added before taking a sample for centrifugation.

The incubation time is chosen from part D (time at which the steady state is reached), the measurements are performed as described in part C, samples from above and below the oil layer are taken again for radioactivity measurements.

Calculations

In order to measure the volume of the mitochondrial matrix, it should be recalled that both membranes are permeable to water, whereas sucrose crosses only the outer membrane.

Mitochondria are incubated in the presence of [^{14}C] sucrose and ^{3}H$_2$O (tubes 13–15) and the matrix volume, V (μl mg^{-1} protein), is calculated according to the following formula:

$$V = \frac{500}{a} \cdot \left(\frac{DPM_H \text{ below oil}}{DPM_H \text{ above oil}} - \frac{DPM_C \text{ below oil}}{DPM_C \text{ above oil}} \right), \qquad (3)$$

where a is the amount of mitochondrial protein (mg) in 500 μl of the centrifuged sample, DPM$_H$ and DPM$_C$ represent the *total* amount of tritium and carbon-14, respectively.

The percentage of radioactive sucrose found below the oil allows to correct for the amount of radioactive substances associated with mitochondria outside their matrix compartment (the so-called carry-down). For any substrate, the amount of substance, Z (nmol mg^{-1} protein), present within the matrix can be calculated in the following way:

$$Z = \frac{DMP_Z \text{ below oil} - \left(\dfrac{DPM_s \text{ below oil}}{DPM_s \text{ above oil}} \cdot DMP_Z \text{ above oil} \right)}{r \cdot a}, \qquad (4)$$

where the indices Z and s refer to the substrate and sucrose, respectively, r is the specific radioactivity of the substrate in DPM $nmol^{-1}$, a is the amount of protein (mg) in 500 μl of the centrifuged sample, DPM below and above the oil represent the total radioactivity of the corresponding fraction.

The amount of the substrate transported, Z, divided by the matrix volume, gives the concentration (mM) of the substrate inside the mitochondria $[c]_{in}$:

$$Z/V = [c]_{in} . \tag{5}$$

In the case of pyruvate, only the 2-cyano-4-hydroxycinnamate-sensitive accumulation should be taken into account. ΔpH and $\Delta \psi$ can be calculated according to the following formulas:

$$\Delta pH = \log \frac{[acetate]_{in}}{[acetate]_{out}} \tag{6}$$

$$\Delta pH \text{ (mV)} = 55.29 \, \Delta pH \text{ (at 6}^{o}\text{C)} \tag{7}$$

$$\Delta \psi \text{ (mV)} = \frac{RT}{F} . \ln \frac{[TPMP^{+}]_{in}}{[TPMP^{+}]_{out}} = 55.29 \cdot \log \frac{[TPMP^{+}]_{in}}{[TPMP^{+}]_{out}} \text{ (at 6}^{o}\text{C)}. \tag{8}$$

The data summarized in the following table are the result of a typical experiment.

Carry down:	1.2 nmol pyruvate mg^{-1} protein
Matrix volume:	0.8 μl mg^{-1} protein
Amount of pyruvate in the matrix after 5 min (Z):	2.3 nmol mg^{-1} protein
$[Pyruvate]_{in}$ (C_{in})·	2.9 mM
$[Pyruvate]_{in}/[pyruvate]_{out}$:	5.76
Pyruvate transport (initial) (2-cyano-4-hydroxycinnamate sensitive):	1.1 nmol min^{-1} mg^{-1} protein
$[Acetate]_{in}/[acetate]_{out}$:	6.1
ΔpH (mV):	43.4 mV
$[TPMP^{+}]_{in}/[TPMP^{+}]_{out}$:	50
$\Delta \psi$:	94 mV
Protonmotive force, Δp :	137 mV

The similarity between pyruvate and acetate transport points to an exchange mechanism between pyruvate and OH^- or a symport between pyruvate and H^+.

References

Bakeeva LE, Grinius LL, Jasaitis AA, Kuliene VV, Levitsky DO, Liebermann EA, Severina II, Skulachev VP (1970) Biochim Biophys Acta 216:113−121

Boyer PD, Chance B, Ernster L, Mitchell P, Racker E. Slater EC (1977) Annu Rev Biochem 46:955−1026

Gornall AG, Bardawill Ch, David MM (1949) J Biol Chem 177:751−766

Halestrap AP (1975) Biochem J 148:85−96

Halestrap AP (1978) Biochem J 172:377:387

Halestrap AP, Denton RM (1975) Biochem J 148:97−106

Halestrap AG, Scott RD, Thomas AP (1980) Int J Biochem 11:97−105

LaNou KF, Schoolwerth AC (1979) Annu Rev Biochem 249:7514−7521

Nalecz KA, Wojtczak AB, Wojtczak L (1984) Biochim Biophys Acta 805:1−11

Nicholls DG (1977) Eur J Biochem 77:349−356

Palmieri F, Klingenberg M (1979) Methods Enzymol 56:279−301

Paradies G, Papa S (1976) FEBS Lett 62:318−321

Paradies G, Papa S (1977) Biochim Biophys Acta 462:333−346

Roos N, Benz R, Brdiczka D (1982) Biochim Biophys Acta 686:204−214